极客学院
jikexueyuan.com

互联网+职业技能系列
职业入门 | 基础知识 | 系统进阶 | 专项提高

HTML5 游戏开发案例教程

Cases Courses of HTML5 Game Development

极客学院 出品

陈惠红 石坤泉 主编 刘世明 谢建华 汤双霞 副主编

人民邮电出版社

北京

图书在版编目（CIP）数据

HTML5游戏开发案例教程 / 陈惠红，石坤泉主编. --
北京：人民邮电出版社，2016.9（2021.1重印）
（互联网+职业技能系列）
ISBN 978-7-115-42672-7

Ⅰ．①H… Ⅱ．①陈… ②石… Ⅲ．①超文本标记语言
－游戏程序－程序设计－教材 Ⅳ．①TP312

中国版本图书馆CIP数据核字(2016)第134627号

内 容 提 要

本书是HTML5游戏开发课程教材。全书分为6章，内容包括构建Canvas开发环境、Canvas基本功能、CreateJS函数库、简单效果案例、HTML5小型游戏、太空英雄大战游戏。全书每章内容与实例紧密结合，并与极客学院网站视频教学课程相结合，学生可以扫描二维码进行视频课程学习，有助于学生随时理解知识、应用知识，使得视频、书籍和课堂紧密配合，达到学以致用的目的。

本书可作为计算机及相关专业的教材，也可作为相关技术人员的参考书或培训教材。

◆ 主　　编　陈惠红　石坤泉
　副 主 编　刘世明　谢建华　汤双霞
　责任编辑　桑　珊
　执行编辑　左仲海
　责任印制　焦志炜

◆ 人民邮电出版社出版发行　北京市丰台区成寿寺路11号
邮编 100164　电子邮件 315@ptpress.com.cn
网址 http://www.ptpress.com.cn
大厂回族自治县聚鑫印刷有限责任公司印刷

◆ 开本：787×1092　1/16
印张：13　　　　　　　　2016年9月第1版
字数：340千字　　　　　　2021年1月河北第5次印刷

定价：35.00元

读者服务热线：(010)81055256　印装质量热线：(010)81055316
反盗版热线：(010)81055315

前言
Foreword

HTML5 自从 2010 年正式推出以来,就以惊人的速度被迅速推广,世界各知名浏览器厂商对 HTML5 都有很好的支持,基础 HTML5 网页端的游戏产品也越来越丰富。如今,只使用 Canvas、JavaScript 和 CreateJS 的几行简单代码就可以在屏幕上画出直线、圆弧、圆和椭圆,还可以指定事件和事件处理来生成动画,并对用户的动作做出响应,可以使用 CreateJS 标准控件在游戏里面加入图片、文字,或者根据需要在游戏中播放视频和音频,设计复杂的游戏。

目前很多高校计算机专业和 IT 培训公司都将 HTML5 游戏设计作为游戏开发课程的必备课之一,而与游戏开发相对应的教材极为稀缺。本书是在作者的教学实践和极客学院技术支持的基础上完成的,全书以学生课堂实践为基础,以案例教学方法为主轴,并结合了极客学院网站视频。学生可以通过扫描二维码的方式直接进入对应视频课程学习。本书一方面跟踪 HTML5 游戏开发技术的发展,适应市场需求,精心选择案例,突出重点、强调实用,使知识讲解全面、系统;另一方面,设计典型案例,将课堂教学与视频网站教学相结合,即有利于学生学习知识,又有利于指导学生实践,真正使得课堂动起来。

本书作为教材使用时,课堂教学建议 35~40 学时,实例教学建议 25~30 学时,教学视频可以通过计算机、手机随时观看。各章主要内容和学时分配如下,老师可以根据实际教学情况进行调整。

章	主要内容	课堂学时	实例学时	视频学时
第 1 章	构建 HTML5 Canvas 游戏开发环境,包括网页游戏概述、游戏开发流程、HTML5 基础知识、开发服务配置、开发工具使用和浏览器选择	2	1	2
第 2 章	Canvas 基本功能,包括 Canvas 标签定义和使用、Canvas 图形、Canvas 文本 Canvas 图片等	6	4	2
第 3 章	CreateJS 函数库,包括初识 CreateJS、CreateJS 包简介、Easel 基础、CreateJS 控件和 Tween 函数包	6	4	2
第 4 章	简单效果案例 帧动画效果 跳舞蝴蝶效果 颜色拼图效果 图像处理效果 跑跳处理效果 炫酷实现效果	7	7	4
第 5 章	围住神经猫游戏的设计与实现	6	4	2
	看你有多色游戏的设计与实现	6	4	2
第 6 章	太空英雄大战游戏的设计与实现	10	8	4

本书由极客学院提供在线课程学习视频，陈惠红、石坤泉担任主编，刘世明、谢建华、汤双霞担任副主编，李玲玲、丘美玲等对本书内容提供了技术指导，陆国甲提供了良好的运行平台，吴晓澜帮助校正了很多格式错误，这里一并感谢。

本书的教学资源可登录人民邮电出版社教育社区（www.ryjiaoyu.com）免费下载，包括本书所需要的软件、所有实例和综合游戏设计源代码，其中源代码经过严格测试，可以在 Windows XP/Windows 7 等平台、Google 浏览器下编译和运行。

由于编者水平有限，书中难免存在疏漏和不足之处，敬请广大读者批评指正，以便本书的后续版本能得到改进和完善。欢迎用书教师加入 QQ 群（539195508），与编者和极客学院视频作者共同探讨交流。

扫一扫加入 QQ 群

编　者
2016 年 4 月

如何使用本书

本套丛书由极客学院精心打造,通过大数据分析,把握企业对职业技能的核心需求,结合极客学院线上课程学习,开启O2O学习新模式。

第1步:创建线上学习账号

使用微信扫描如下二维码,自动创建(登录)极客学院账号,并自动加入与本书配套的线上社群。

第2步:立体化学习

创建账号后,即可开始学习,除了学习图文内容外,还可以扫描书中二维码观看配套视频课程,下载对应资料,查看常见问题并提问,参与社群讨论。

资料　　　　　视频　　　　　问题

第3步:学习结果测评

完成学习后,可以扫描以上二维码,参加本书测评,成绩合格者可以申请课程结业证书,成绩优秀者将会获得额外大奖。

目录
Contents

开发准备篇

第1章 构建 Canvas 开发环境   2
- 1.1 网页游戏概述   3
- 1.2 游戏开发流程   4
- 1.3 HTML5 基础知识   5
  - 1.3.1 HTML5 概述   5
  - 1.3.2 Canvas 简介   6
- 1.4 开发环境配置   7
  - 1.4.1 开发服务器   7
  - 1.4.2 开发工具   7
  - 1.4.3 浏览器   13

基础知识篇

第2章 Canvas 基本功能   16
- 2.1 Canvas 标签   17
  - 2.1.1 定义 Canvas 标签   17
  - 2.1.2 理解 Canvas 坐标系   18
  - 2.1.3 获取 Canvas 环境上下文   19
- 2.2 Canvas 图形   20
  - 2.2.1 绘制 Canvas 路径   20
  - 2.2.2 绘制 Canvas 变形图形   31
  - 2.2.3 处理 Canvas 图形   40
- 2.3 Canvas 文本   45
  - 2.3.1 绘制文字   45
  - 2.3.2 设置文字格式   47
  - 2.3.3 设置文字对齐方式   52
- 2.4 Canvas 图片   55
  - 2.4.1 绘制 drawImage 图片   55
  - 2.4.2 使用 getImageData()和 putImageData()绘制图片   57
  - 2.4.3 使用 createImageData()新建像素   59

第3章 CreateJS 函数库   62
- 3.1 初识 CreateJS   63
  - 3.1.1 下载 CreateJS   63
  - 3.1.2 介绍 CreateJS   64
  - 3.1.3 对比 CreateJS 与 Canvas   65
- 3.2 CreateJS 包简介   69
  - 3.2.1 EaselJS 包   69
  - 3.2.2 TweenJS 包   70
  - 3.2.3 SoundJS 包   72
  - 3.2.4 PreLoadJS 包   73
- 3.3 EaselJS 基础   75
  - 3.3.1 EaselJS 容器   75
  - 3.3.2 EaselJS 绘图   82
  - 3.3.3 EaselJS 事件   92
- 3.4 CreateJS 控件   101
  - 3.4.1 Text   101
  - 3.4.2 BitMap   103
  - 3.4.3 MovieClip   106
  - 3.4.4 Sprite   108
  - 3.4.5 DOMElement   111
- 3.5 Tween 函数包   114
  - 3.5.1 CSSPlugin   114
  - 3.5.2 Ease   115
  - 3.5.3 MotionGuidePlugin   117
  - 3.5.4 Tween   119

案例实战篇

第4章 简单效果案例   122
- 4.1 帧动画效果   123
- 4.2 跳舞蝴蝶效果   125

4.3	颜色拼图游戏	127
4.4	图像处理效果	130
4.5	处理跑跳效果	134
4.6	实现炫酷效果	139

第 5 章 HTML5 小型游戏　　142

5.1	围住神经猫游戏	143
5.1.1	介绍围住神经猫游戏的玩法	143
5.1.2	使用 CreateJS 围住神经猫	145
5.1.3	绘制围住神经猫游戏页面元素	146
5.1.4	添加围住神经猫游戏监听事件	150
5.1.5	使用简单的逻辑实现围住神经猫游戏效果	151
5.1.6	实现围住神经猫游戏完整效果	153
5.2	看你有多色游戏	159
5.2.1	介绍看你有多色游戏	159
5.2.2	使用 CreateJS 开发看你有多色	161
5.2.3	制作看你有多色游戏	162

第 6 章 HTML5 大型游戏——太空英雄大战　　166

6.1	游戏简介	167
6.2	准备项目	167
6.2.1	设置 HTML 文件	167
6.2.2	Sprite Sheet 文件	169
6.2.3	资源管理	171
6.2.4	创建应用类	171
6.3	创建 Sprites	174
6.3.1	创建英雄飞船	174
6.3.2	创建敌方飞船	176
6.3.3	创建子弹和爆炸效果	177
6.4	创建参谋中心（HUD）	179
6.4.1	创建 HUD Sprite 框架	179
6.4.2	创建 HUD 效果图	179
6.4.3	创建分数板	181
6.4.4	创建生命箱	182
6.5	创建对象池	183
6.6	创建场景	184
6.6.1	创建游戏菜单场景	184
6.6.2	创建游戏场景	185
6.6.3	创建游戏结束场景	186
6.7	创建游戏	188
6.7.1	设置游戏参数	188
6.7.2	初始化游戏	189
6.7.3	创建游戏 Sprites	189
6.7.4	设计游戏控制	190
6.7.5	创建游戏循环	191
6.7.6	设置游戏更新功能	192
6.7.7	创建渲染函数	194
6.7.8	创建场景响应函数	195
6.7.9	检测碰撞效果	196
6.7.10	创建检测函数	197

开发准备篇

第1章

构建Canvas开发环境

■ 随着新一代 Web 开发标准——HTML5 的诞生，各大浏览器厂商和软件厂商都不遗余力地支持 HTML5 的发展，加入 HTML5 的阵营，互联网时代的新一轮革命即将展开，当游戏碰上 HTML5 会产生什么样的激烈火花，真是让人期待。

作为本书的第一部分，先来学习一下什么是网页游戏、什么是HTML5、HTML5 的开发和运行环境等。对于游戏开发者来说，游戏的开发需要以面向对象的思维去设计，游戏的设计也遵循一系列的开发流程，所以本章也对游戏的开发流程做一定的说明。

极客学院在线视频学习网址：
http://www.jikexueyuan.com/course/181_1.html
手机扫描二维码

HTML5 开发前准备

1.1 网页游戏概述

从时间上来计算，游戏行业从诞生到现在还不到 100 年的历史，跟其他传统的行业相比，它甚至像襁褓中的婴儿一样小，但正是这个婴儿，正逐渐挑战着众多传统行业。现在，很多人都会在不同的时刻玩着不同的游戏，也许你正在虚拟的网络游戏中热血澎湃地战斗，也许你正在电子游戏竞技中展现你的人生价值，也许你正在忙碌的学习工作之余，玩着切水果的游戏不停地发泄，总之，你会感觉到，它正在悄然融入到生活中，成为生活的一部分。

按照电子游戏的载体来划分，电子游戏现在基本分为 3 个主要的阵营。第一部分是以电视游戏为主，第二部分是以个人电脑游戏为主，第三部分是以手机和平板为主。而从游戏的玩家数量来说，游戏经历了从单机游戏时代到现在的网络游戏时代。随着互联网的普及以及电脑硬件的飞速发展，互联网游戏正处于高速发展的时期，特别是网页游戏得到了空前的发展。

网页游戏（Web Game）是一种无端网游，和《魔兽》系列、《星际》系列等传统的游戏相比，网页游戏不需要下载客户端，玩家只需要通过电脑打开浏览器即可进行游戏，与传统的大型网游比起来，其优点是无需安装、占据空间小、使用方便等，对于开发人员来说，比开发传统的网络游戏更容易。

网页游戏从最早的多用户虚拟空间游戏（Multiple User Domain，MUD）发展而来，玩家爱称"泥巴游戏"。早期的 MUD 游戏限于技术条件，几乎是纯文字网游，没有图形，全部用文字和字符画来构成。按照维基百科记载，世界上第一款真正意义上的实时多人交互网络 MUD 游戏"MUD1"，是在 1978 年由英国埃塞克斯大学的罗伊·特鲁布肖用 DEC-10 编写的。随着 Internet 和 HTML 语言的飞速发展，纯文字类的游戏退出历史舞台，丰富多彩的带图像的网页游戏逐渐兴起。现在的一些 2D 网页游戏几乎能与传统的网络游戏媲美，比如"可乐吧""弹弹堂""第七城市"、4299 游戏平台、91wan 游戏平台、1wan 游戏平台等。

由于网页游戏运行的环境是浏览器，所以常用的开发语言为 HTML、CSS 样式以及 JavaScript 语言，服务器的开发可以使用诸如 C/C++、C#、Java、PHP 等传统的服务器端语言。在 HTML4 时代，HTML 语言受到诸如缺乏高效的图形渲染方法、缺乏实时的网络通信方法等技术支持的限制，加上 JavaScript 运行效率相对于一些常用的游戏编程语言 C/C++、Java 低，所以目前比较成熟的网页技术都需要在浏览器中安装一些特殊的插件（Flash Player、Applet、ActiveX、Unity Web Player 等）以帮助 Web Game 高效运行。就目前来说，Web Game 使用最广泛的客户端技术主要还是以 Flash 平台为主。从 1995 年到现在经过了近 20 多年的时间，各种关于动画、游戏方面的技术已经非常成熟，所以 Flash 通常作为 Web Game 首选开发平台。但随着 HTML 新标准的发布，也就是 HTML5 的横空出世，也就注定了 Flash 的路将不会长久。Flash 的研发公司 Adobe 已经于 2011 年宣布停止 Flash 后续研发工作，而转向新的 HTML5。

HTML5 被看作 Web 开发者创建流行 Web 应用的利器，增加了对视频和 Canvas 2D 的支持。HTML5 的优点主要在于可以进行跨平台的使用。比如开发了一款 HTML5 的游戏，可以很轻易地移植到 UC 的开放平台、Opera 的游戏中心、Facebook 应用平台，甚至可以通过封装的技术发布到 App Store 或 Google Play 上，所以它的跨平台性非常强大，这也是大多数人对 HTML5 感兴趣的主要原因。

1.2 游戏开发流程

人体的生理循环系统被称为生命之河。我们从外界吸收足够的营养和能量，在人的生理循环系统作用下，这些能量和营养提供给身体的各个功能器官进行运作，最后，排除一些并不需要的杂质。正是由于日复一日、年复一年的不断的生理循环，我们的生命才得以延续。正如人的循环系统是生命之河一样，游戏中的主循环也是整个游戏运行的核心环节。

从单机游戏到复杂的大型多人在线游戏，游戏中的主循环部分都包含着类似的几个部分，正是有了这些部分，才有了丰富多彩的游戏世界。图 1-1 所示为游戏基本的运行机制和核心流程。

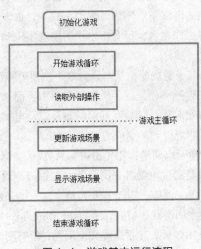

图 1-1 游戏基本运行流程

实际上，游戏的运行就是整个游戏循环的运行，在整个游戏循环过程中需要完成图 1-1 中的主要部分。

1. 初始化游戏

在整个游戏的开始部分，需要做一些游戏初始化的工作，以便游戏更加快速地运行，比如需要加载游戏运行的各种资源文件，读取游戏配置的各项数据等。

2. 游戏主循环部分

（1）读取外部操作。整个游戏的运行离不开玩家和游戏世界的实时交互，甚至来自于游戏世界本身的交互，所以游戏世界中必须监控来自于外部的操作，以随时改变游戏的状态。这些外部操作主要来自外部设备或者网络数据，比如键盘、摇杆、触屏、话筒和方向盘等。

（2）更新游戏场景。在获取了游戏外设的操作后，游戏世界必须根据这些改变游戏世界的数据进行更新整个游戏的场景。例如，在"超级玛丽"中按下了跳跃按钮，这时候就必须在游戏循环中根据外部操作更新玛丽的动作，让它跳得更高。事实上，整个更新的过程并不一定需要等到外来的操作，有可能是由内部游戏的其他对象引发的，比如"超级玛丽"中玛丽不小心碰到了怪物，也有可能是来自于网络游戏的服务器，在网络游戏中，客户端的游戏世界还需要根据服务器的数据进行更新。事实上，这部分也可以看出游戏的主要逻辑处理部分，一旦满足退出游戏逻辑，那么游戏主循环就结束，从而结束游戏。

（3）显示游戏场景。为了让用户有成就感，让用户有更多的视觉享受，在进行了游戏更新场景的操作之后，需要把游戏中的场景显示出来，不管是战神中的血腥，还是仙剑中的凄美动人爱情，最后都需要通过画面渲染出来。

3. 结束游戏循环

游戏主循环中，如果满足退出游戏逻辑，则游戏结束。

1.3 HTML5 基础知识

1.3.1 HTML5 概述

自从 HTML5 新标准发布以来，就引起了互联网技术的新一轮风暴，作为新一代的 Web 技术领航者，它受到了各大厂商的追捧，几乎所有的 IT 大厂商都全力提供对 HTML5 规范的支持。相对于 HTML 4.X 版本而言，HTML5 提供了许多令人激动的新特性，这些新特性将为 HTML5 开创新的 Web 时代提供了坚强的基石。

超文本标记语言（Hypertext Makeup Language，HTML）是专门在 Internet 上传输多媒体的一种语言，正是有了 HTML 语言的出现，现在的互联网世界才显得丰富多彩，从 1993 年第一个版本的 HTML 语言诞生以来，共经历了以下几个重要的发布版本。

（1）HTML（第一版），这是一个非正式的工作版本，于 1993 年 6 月作为 IEIF（Internet Engineering Task Force）草案发布；

（2）HTML 2.0，1995 年 11 月作为 RFC1866（Request For Comment）发布，RFC 是由 IETF 发布的备忘录；

（3）HTML 3.2，1997 年 1 月 14 日成为 W3C（World Wide Web Consortium）推荐标准；

（4）HTML 4.0，1997 年 12 月 18 日，W3C 推荐标准；

（5）HTML 4.01，1999 年 12 月 24 日，W3C 推荐标准；

（6）Web Application 1.0，2004 年作为 HTML5 草案的前身由 WHATWG（Web Hypertext Application Technology Working Group，为推动 HTML5 标准而建立的组织）提出，2007 年被 W3C 组织作为推荐标准。

（7）HTML5 草案，2008 年 1 月 22 日，第一份草案正式发布。

（8）HTML 5.1，2012 年 12 月 17 日，W3C 的首份规范草案发布。

事实上到现在为止，HTML5 还处于发展和完善时期，但诸多 HTML5 中新增加的功能已经受到各大软件厂商的鼎力支持。从 HTML5 前身的名称（Web Application）可以看出 HTML5 的决心，HTML 不再只是单纯的网站制作语言，而是作为 Web 应用程序的开发语言应运而生。为了能够承担 Web 应用程序所能够完成的功能，在不需要安装任何插件的情况下，HTML5 中提供了以下激动人心的功能。

（1）Canvas 画布元素。Canvas 元素的诞生为 HTML5 能够支持较高性能的动画和游戏提供了条件。Canvas 可以直接使用硬件加速完成像素级别的图像渲染，不仅可以完成 2D 图形渲染，使用 WebGL 以及 Shader 语言还可以完成较高性能的 3D 图形渲染。

（2）多媒体元素。HTML5 中提供了专门的 Audio 元素和 Video 元素，用于播放网络音频文件和视频文件，有了这两个多媒体元素，不再需要单独安装插件就可以进行影音的播放，减少了浏览器的污染程序。

（3）地理信息服务。通过 HTML5 的地理信息服务 API 可以获取到客户端所在的经度和纬度，利用这些信息可以向这个坐标附近的区域提供服务，可应用于地理交通信息查询、基于 LBS（Location Based Services）的社交游戏等。

（4）本地存储服务。相对于传统的 Cookie 微量的本地存储技术，HTML5 推出了新的本地存储规范，提供了容量更大、更安全和更易于使用的本地存储方案。

（5）WebSocket 通信。弥补了传统 Web 应用程序缺乏实时通信功能的不足，使用 WebSocket 技术可以在 Web 应用程序中实现类似于传统 C/S 结构应用程序的通信功能，使得在 Web 环境中构建实时的通信程序成为可能。

（6）离线存储。HTML5 的离线缓存应用的功能，使客户端即使没有连接到互联网络的情况下，也可以在客户端正常使用本地功能。有了这个强大的功能，用户可以更加灵活地控制缓存资源的加载，可以在没有网络信号的情况下使用本地应用。

（7）多线程。HTML5 中提供了真正意义上的多线程解决方案，在 HTML4 的使用过程中，如果遇到客户端需要在后台执行耗时方法，则页面会处于"假死"状态，而在 HTML5 中可以使用多线程解决类似问题。

（8）设备。为了能够适应多种设备（PC、手机和平板），HTML5 提供了 Device 元素，可以让应用程序访问诸如摄像头、麦克风等硬件设备。

总之，这些新增的特性无疑都是冲着本地应用程序而来，尽管 HTML5 还处于发展阶段，但已经成为下一代 Web 开发的标准。

1.3.2 Canvas 简介

在 HTML5 的王国中，将使用具有魔力的 Canvas 元素，来在浏览器中做一番奇妙的事情。如图 1-2 所示的图像查看器、一个交互式的放大镜、一个可以在各种类型的浏览器以及 iPad 之中运行的运动效果程序，还有完整的 HTML5 游戏，以前这些软件都是在 Flash 开发领域中实现的。

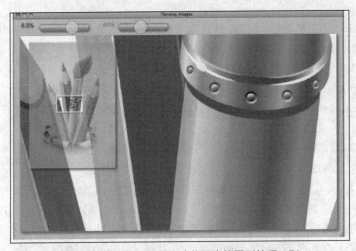

图 1-2　Canvas 提供了功能强大的图形处理 API

HTML5 <canvas>标签用于绘制图像（通过脚本，通常是 JavaScript）。不过，<canvas>元素本身并没有绘制能力（它仅仅是图形的容器），必须使用脚本来完成实际的绘图任务。getContext()方法可返回一个对象，该对象提供了用于在画布上绘图的方法和属性。本节概要说明几个重要的 getContext("2d")对象的属性和方法，具体可参照之后几节在画布上绘制文本、线条、矩形、圆形的使用方法。其他的属性和方法请参照教学资源的参考手册。

1.4 开发环境配置

HTML5 有很多新的特性，各个浏览器对这些特性的支持度也都不一样。因为本书介绍的是基于 Canvas 的开发，并且有一部分的开发效果只有在服务器才能显示出来，所以在这里只介绍对 Canvas 的支持情况。

1.4.1 开发服务器

对于 Mac 来说，苹果系统本身自带本地服务器，对于 Windows 来说，则推荐使用 XAMPP（Apache+MySQL+PHP+PERL），它是一个功能强大的服务器系统开发套件，可以在 Windows、Linux、Solaris 等操作系统下安装使用，支持多语言（如英语、简体中文、繁体中文、韩文、俄文和日文等）。

XAMPP 的官方网址为：http://www.apachefriends.org/。

下载安装后，打开 xampp 文件夹并启动 xampp-control.exe 文件，然后单击 Tomcat 右侧的【Start】按钮启动 Tomcat，Tomcat 启动后状态变为"Running"，如图 1-3 所示。

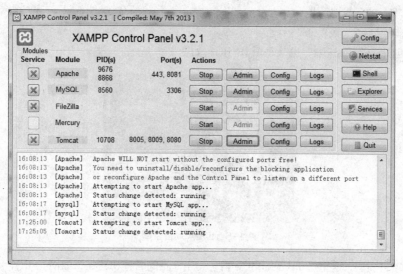

图 1-3　XAMPP 启动页面

1.4.2 开发工具

原则上来说，使用任何文本编辑工具都可以完成 HTML5 代码的编写工作，编辑好 HTML5 代码保存为 .htm 或者 .html 文件即可，然后可以使用支持 HTML5 的浏览器查看效果。

工欲善其事必先利其器，尽管可以直接使用 NotePad 编写 HTML5 应用程序，但为了提高代码的编写效率和减少出错概率，可以使用一些比较常用的 IDE 工具完成相关程序开发，这里提供了几个 IDE 工具，本书主要使用 IntelliJ IDEA 工具来开发和管理 HTML5 项目。

1. Adobe Dreamweaver CS6

Dreamweaver CS6 是世界顶级软件厂商 Adobe 推出的一套拥有可视化编辑界面，用于制作并编辑网站和移动应用程序的网页设计软件。由于 Dreamweaver 支持代码、拆分、设计、实时视图等多种方式来创作、编写和修改网页，对于初级人员，可以无需编写任何代码就能快速创建 Web 页面。如图 1-4 所示，其成熟的代码编辑工具更适用于高级 Web 开发人员的创作。

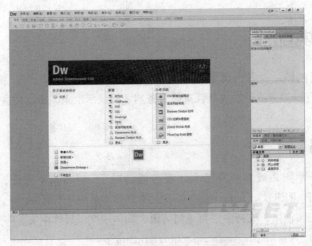

图 1-4 Dreamweaver 界面图

2. IntelliJ IDEA

IntelliJ IDEA 的下载地址是 http://www.jetbrains.com/idea/download/，下载安装后打开 IDEA，下面开始一步步地创建一个 Web 项目。

（1）首先创建一个 Project，也就是项目空间，如图 1-5 所示。

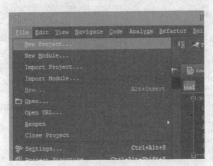

图 1-5 IDEA 项目空间

（2）选择项目类型，这里选 Java Module 自定义工作空间名称和路径，如图 1-6 所示。

图 1-6 IDEA 项目空间类型

（3）选择需要用到的框架组件，这里只选了第一个 Web Application -> Finish，如图 1-7 所示。

图 1-7　IDEA 项目框架组件

（4）创建完工作空间，如图 1-8 所示，默认会是一个 Module 也就是一个项目，但是不推荐使用该项目进行开发。

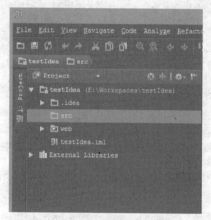

图 1-8　IDEA 项目默认 Module

（5）在该项目空间中，添加新的工程，选中工作空间，右键单击 Open Module Settings 或者按下<F4>键，如图 1-9 所示。

图 1-9　IDEA 项目空间添加新工程菜单

（6）添加工程，如图 1-10 所示。

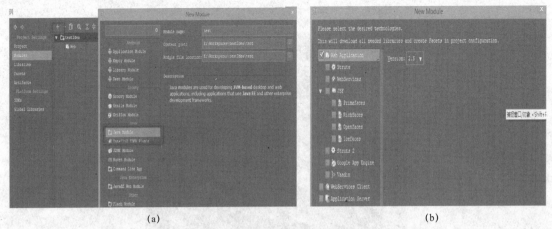

图 1-10　IDEA 项目添加工程 1

（7）完成添加项目后，可以在新建工程的 web>WEB-INF 下创建 classes 和 lib 文件夹，如图 1-11 所示。

图 1-11　IDEA 项目创建 lib 和 classs 文件夹

（8）修改编译输出目录，将 Paths>Use module compile output path 转到自定义的 classes 文件夹，如图 1-12 所示。

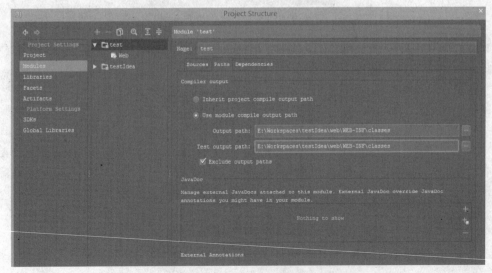

图 1-12　IDEA 项目修改输出目录

（9）同样可以指定 lib 库目录，添加>jars or directories 指向创建的 lib 文件夹，弹出窗口选择 jar directory，如图 1-13 所示。

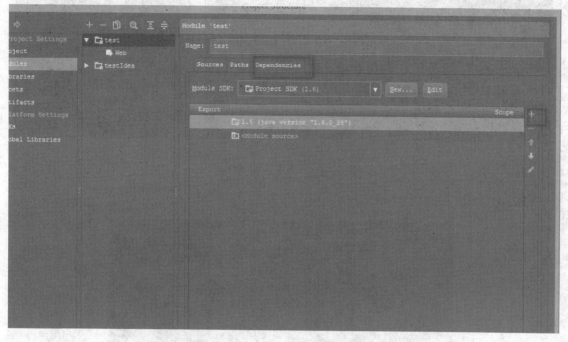

图 1-13　IDEA 项目制定 lib 库目录

（10）点击图 1-14（a）所示的 Edit Configuration，进行进入 Tomcat 配置页面，进入页面后，点击右侧的"+"弹出页面后，按照图 1-14（b）所示选择 Tomcat Server>Local。

(a)

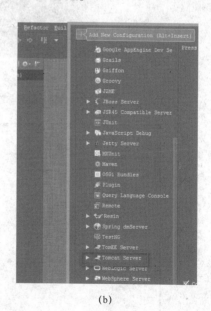

(b)

图 1-14　IDEA 项目部署 tomcat server

（11）选择 tomcat 的部署的应用 deployment，添加一个 deployment，如图 1-15 所示。

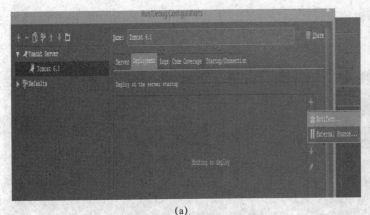

(a)   (b)

图 1-15   IDEA 项目部署应用

（12）启动测试，如图 1-16 所示。

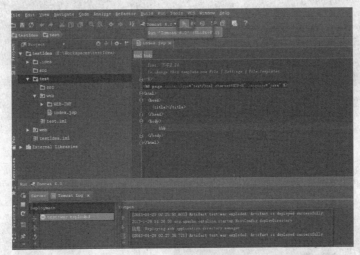

图 1-16   IDEA 项目启动测试

（13）测试完成，测试后出现的网页效果如图 1-17 所示。

图 1-17   IDEA 项目测试结果显示

1.4.3 浏览器

HTML5 仍处于完善之中，然而，现在大部分浏览器已经开始具备对 HTML5 的支持。当然，各大浏览器的开发还在继续，将来应该会全面支持 HTML5 的。浏览器厂商的竞争促使各大浏览器对 HTML5 和 CSS3 的支持越来越完善，图 1-18 列出了 IE、Chrome、Firefox、Safari、Opera 五大主流浏览器对内置 Canvas、Audio、Video、SVG 和 WebGL 等重要特性对象的支持情况。

Chrome、Firefox、Safari 和 Opera 支持全部的特性，其中对于 WebGL，IE9 是不支持的，Firefox、Safari 和 Opera 都是部分支持，只有 Chrome 完全支持。

平台	MAC				WIN						
浏览器	CHROME	FIREFOX	OPERA	SAFARI	CHROME	FIREFOX	OPERA	SAFARI	IE		
版本	5	3.6	10.1	4	4	3.6	10	10.5	4	6	7	8
Canvas	✓	✓	✓	✓	✓	✓	✓	✓	✓	✗	✗	✗
Canvas Text	✓	✓	✗	✓	✓	✓	✗	✗	✓	✗	✗	✗
Audio	✓	✓	✗	✓	✓	✓	✓	✗	✓	✗	✗	✗
Video	✓	✓	✗	✓	✓	✓	✗	✗	✓	✗	✗	✗

图 1-18 主流浏览器支持情况

Html5test.com 网站主要针对 HTML5 进行兼容性测试，以 Chrome 为例，其 HTML5 测试评分如图 1-19 所示。

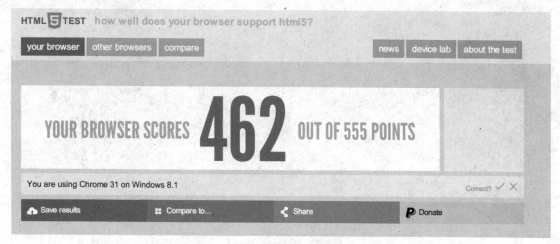

图 1-19 浏览器兼容性测试结果

对 HTML5 综合性能进行检测的权威网站 http://peacekeeper.futuremark.com 可以针对浏览器进行全方位的测试，将 Chrome 浏览器缓存清理完毕之后，关闭计算机系统中所有第三方进程，运行 Chrome 浏览器，并将 Peacekeeper 测试地址输入到 Chrome 地址栏。页面加载完毕之后，是 Peacekeeper 为 Chrome 浏览器的测试准备页面，点击 "GO" 测试开始。Peacekeeper 在正式测试开始之前，会自动检测所处的浏览器平台，并在页面中呈现浏览器受关注的排列情况。直接点击 "浏览器衡量" 正式开始测试 Chrome 浏览器。

经过多个测试关键页面元素的加载、呈现等运作，Peacekeeper 需要 5~8 分钟的时间来运行各项测试指标。如图1-20所示,最终所得到的测试结果为1720。从检测结果来看,目前综合性能最高的也是Chrome浏览器,当然每个浏览器的好坏不是仅凭这个就可以定义的,并且即使浏览器的功能再强大,界面再漂亮,也不一定就是用户心目中最好的浏览器,因为浏览器的使用涉及一个习惯问题,用习惯了自然就觉得好了,根本不会去考虑它功能是否强大。另外现实生活中使用的浏览器,虽然其功能十分强大,十分完善,但是并不是每个人都能完全用到所有的功能。

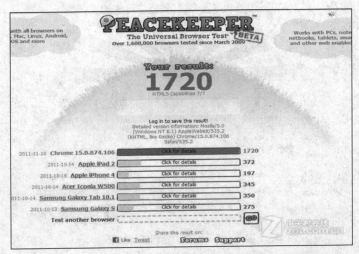

图1-20　Chrome浏览器Peacekeeper beta版测试结果

基础知识篇

第2章

Canvas基本功能

■ 在所有的桌面应用程序的开发平台中几乎都有 Canvas 组件。Canvas 组件已经成为绘图组件的代名词。Canvas 元素本质上是在浏览器中提供了一块可绘制的区域，JavaScript 代码可以访问该区域，通过一套完成的 API 进行二维图像绘制。另外，如果显卡硬件支持 3D 图形功能，还可以使用 Canvas 绘制 3D 图形。本章主要讲解 HTML5 Canvas 的基本功能，利用 Canvas 的 API，用实例展示一些基本图形的绘制及操作方法，包括 Canvas 图形、文本和图像的操作等。

2.1 Canvas 标签

极客学院在线视频学习网址：
http://www.jikexueyuan.com/course/221_1.html
手机扫描二维码

HTML5 Canvas 标签-创建 Canvas

2.1.1 定义 Canvas 标签

<canvas>标签定义图形，比如图表和其他图像，必须使用脚本来绘制图形。如图 2-1 所示，在画布（Canvas）上画一个红色矩形、渐变矩形、彩色矩形和一些彩色的文字。

图 2-1 Canvas 标签定义图形效果

一个画布在网页中是一个矩形框，通过<canvas>元素来绘制。

【案例 2-1】创建一个画布（Canvas）

代码清单 2-1

```
<!DOCTYPE html>
<html>
<body>
<canvas id="myCanvas" width="200" height="100" style="border:1px solid #000000;">
您的浏览器不支持 HTML5 canvas 标签。
</canvas>
</body>
</html>
```

默认情况下<canvas>元素没有边框和内容，可以使用 style 属性来添加边框。标签通常需要指定一个 id 属性（脚本中经常引用），width 和 height 属性定义画布的大小。

 可以在 HTML 页面中使用多个<canvas>元素。

程序运行效果如图 2-2 所示。

图 2-2　代码清单 2-1 运行后的效果图

2.1.2　理解 Canvas 坐标系

如图 2-3 所示，Canvas 元素的坐标系是一个二维网格，它的坐标系是以左上角为原点，向右延伸是横坐标 x 的正方向，向下延伸是纵坐标 y 的正方向，所以原点的坐标是$(x,y)=(0,0)$，弄清楚坐标系对于以后使用 Canvas 的 API 意义重大，在绘图的时候需要时刻记住坐标空间。

图 2-3　Canvas 坐标系

例如第 2.2 节的 fillRect 方法拥有参数 $(0,0,150,75)$，意思是：在画布上绘制 150×75 的矩形，从左上角(0,0)开始。

【案例 2-2】使用 JavaScript 来获取 Canvas 坐标

如代码清单 2-2 和图 2-3 所示，画布的 x 和 y 坐标用于在画布上对绘画进行定位。当鼠标移动到矩形框上时，显示定位坐标。

代码清单 2-2

```
<div id="coordiv" style="float:left;width:199px;height:99px;border:1px solid #c3c3c3" onmousemove="cnvs_getCoordinates(event)" onmouseout="cnvs_clearCoordinates()"></div>
<div id="xycoordinates"></div>

<script>
function cnvs_getCoordinates(e)
{
x=e.clientX;
y=e.clientY;
document.getElementById("xycoordinates").innerHTML="Coordinates: (" + x + "," + y + ")";
}

function cnvs_clearCoordinates()
{
document.getElementById("xycoordinates").innerHTML="";
}
</script>
```

程序运行效果如图 2-4 所示。

图 2-4　代码清单 2-2 运行后的效果图

2.1.3　获取 Canvas 环境上下文

在定义好 Canvas 之后，就可以使用 JavaScript 访问 Canvas 元素，使用 Canvas 提供的一系列 API。在使用 Canvas 时，首先要得到 Canvas 的环境上下文，才能够对 Canvas 进行相应操作，可以通过【案例 2-3】获取环境上下文。

【案例 2-3】获取 Canvas 的环境上下文

代码清单 2-3

```
<!DOCTYPE html>
<html>
<body>
<canvas id="myCanvas" width="200" height="100" style="border:1px solid #c3c3c3;">
您的浏览器不支持 HTML5 canvas 标签。
</canvas>

<script>
var c=document.getElementById("myCanvas");
var ctx=c.getContext("2d");
ctx.fillStyle="#FF0000";
ctx.fillRect(0,0,150,75);
</script>
</body>
</html>
```

❖　代码解析：

Canvas 元素本身是没有绘图能力的。所有的绘制工作必须在 JavaScript 内部完成。

首先，找到<canvas>元素：

```
var c=document.getElementById("myCanvas");
```

然后，创建 context 对象：

```
var ctx=c.getContext("2d");
```

getContext("2d") 对象是内建的 HTML5 对象，拥有多种绘制路径、矩形、圆形、字符以及添加图像的方法。

下面的两行代码绘制一个红色的矩形：

```
ctx.fillStyle="#FF0000";
ctx.fillRect(0,0,150,75);
```

设置 fillStyle 属性可以是 CSS 颜色、渐变或图案。fillStyle 默认设置是#000000（黑色）。

fillRect(x,y,width,height)方法定义了矩形当前的填充方式。

程序运行效果如图 2-5 所示。

图 2-5　代码清单 2-3 运行后的效果图

2.2　Canvas 图形

创建 Canvas 和获取了 Canvas 环境上下文之后，就可以开始进行绘图了，绘图的方式有两类：一类是进行图形绘制，另一类是图形的处理。

极客学院在线视频学习网址：
http://www.jikexueyuan.com/course/221_2.html
手机扫描二维码

HTML5 Canvas 标签-绘制图形

2.2.1　绘制 Canvas 路径

所谓基本图形，就是指线、矩形、圆等最简单的图形，任何复杂的图形都是由这些简单的图形组合而成的。

1．绘制线条

在 Canvas 上画线，将使用以下两种方法：moveTo(x,y) 定义线条开始坐标；lineTo(x,y) 定义线条结束坐标。

绘制线条必须使用到"ink"的方法，就像 stroke()。

【案例 2-4】用 moveTo() 绘制 Canvas 线条

极客学院在线视频学习网址：
http://www.jikexueyuan.com/course/1299_2.html
手机扫描二维码

HTML5 moveTo 与 lineTo

代码清单 2-4

```
<script>
var c=document.getElementById("myCanvas");
var ctx=c.getContext("2d");
ctx.moveTo(0,0);
ctx.lineTo(200,100);
ctx.stroke();
</script>
```

❖ 代码解析

定义开始坐标(0,0)和结束坐标(200,100)。然后使用 stroke()方法来绘制线条,代码运行效果如图 2-6 所示。

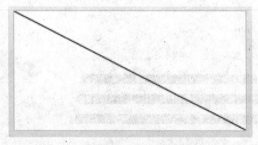

图 2-6　代码清单 2-4 运行后的效果图

【案例 2-5】绘制圆形的结束线帽

代码清单 2-5

```
<script>
var c=document.getElementById("myCanvas");
var ctx=c.getContext("2d");
ctx.beginPath();
ctx.lineWidth=10;
ctx.lineCap="butt";
ctx.moveTo(20,20);
ctx.lineTo(200,20);
ctx.stroke();

ctx.beginPath();
ctx.lineCap="round";
ctx.moveTo(20,40);
ctx.lineTo(200,40);
ctx.stroke();

ctx.beginPath();
ctx.lineCap="square";
ctx.moveTo(20,60);
ctx.lineTo(200,60);
ctx.stroke();
</script>
```

❖ 代码解析

lineCap 属性设置或返回线条末端线帽的样式。其对应的语法格式如表 2-1 所示。

表2-1 lineCap 语法

默认值	butt
JavaScript 语法	*context*.lineCap="butt|round|square";

属性值如表 2-2 所示。

表2-2 lineCap 属性值

属性值	描述
butt	默认。向线条的每个末端添加平直的边缘
round	向线条的每个末端添加圆形线帽
square	向线条的每个末端添加正方形线帽

程序运行效果如图 2-7 所示。

图 2-7 代码清单 2-5 运行后的效果图

【案例 2-6】使用 moveTo（ ）与 lineTo（ ）绘制复杂图形

代码清单 2-6

```
<!DOCTYPE html>
<html>
<head lang="en">
<meta charset="utf-8">
<title></title>
<script>
function draw(id){
    var canvas = document.getElementById(id);
    var context = canvas.getContext("2d");
        context.fillStyle = "#eeeeef"; //设置绘图区域颜色
        context.fillRect(0,0,300,400); //画矩形
    var dx = 150;
    var dy = 150;
    var s =100;
        context.beginPath();//开始绘图
        context.fillStyle = "rgb(100,255,100)"; //设置绘图区域颜色
        context.strokeStyle = "rgb(0,0,100)";//设置线条颜色
    var x = Math.sin(0);
    var y = Math.coas(0);
```

```
        var dig = Math.PI / 15*11;
          for(var i = 0; i<30; i++){    //不断地旋转绘制线条
            var x = Math.sin(i*dig);
            var y = Math.cos(i*dig);
            context.LineTo(dx+x*s,dy+y*s);
          }
        context.closePath();
        context.fill();
        context.stroke();
      }
</script>
</head>
<body onload="draw('canvas')">
<!--move to line to-->
<canvas id="canvas" width="300" height="400"></canvas>
</body>
</html>
```

◆ 代码解析

本案例利用循环与 moveTo 与 lineTo 形成复杂结果,主要是将光标移动到指定坐标点,绘制直线的时候以这个坐标点为起点,用 moveTo(x, y)画图到 x、y 轴的位置,案例运行效果如图 2-8 所示。

图 2-8 代码清单 2-6 运行后的效果图

2. 绘制矩形

【案例 2-7】使用 rect()与 stroke()绘制 Canvas 矩形

代码清单 2-7

```
<script>
var c=document.getElementById("myCanvas");
var ctx=c.getContext("2d");
ctx.rect(20,20,150,100);
ctx.stroke();
</script>
```

◆ 代码解析

本案例主要使用 stroke()或 rect()方法在画布上实际地绘制矩形。其对应的语法格式如表 2-3 所示。

表 2-3　rect()语法

JavaScript 语法	*context*.rect(*x*, *y*, *width*, *height*);

参数值如表 2-4 所示。

表 2-4　rect()参数

参数	描述
x	矩形左上角的 *x* 坐标
y	矩形左上角的 *y* 坐标
width	矩形的宽度，以像素计
height	矩形的高度，以像素计

程序运行效果如图 2-9 所示。

【案例 2-8】使用 fillRect()绘制 Canvas 矩形

代码清单 2-8

```
<script>
var c=document.getElementById("myCanvas");
var ctx=c.getContext("2d");
ctx.fillRect(20,20,150,100);
</script>
```

◆ 代码解析

本案例使用 fillRect()方法绘制"已填充"的矩形。默认的填充颜色是黑色，对应的语法格式如表 2-5 所示。

可以使用 fillStyle 属性来设置用于填充绘图的颜色、渐变或模式。

表 2-5　fillRect()语法

JavaScript 语法	*context*.fillRect(*x*, *y*, *width*, *height*);

参数值如表 2-6 所示。

表 2-6　fillRect()参数

参数	描述
x	矩形左上角的 *x* 坐标
y	矩形左上角的 *y* 坐标
width	矩形的宽度，以像素计
height	矩形的高度，以像素计

程序运行效果如图 2-9 所示。

图 2-9 代码清单 2-7 和代码清单 2-8 运行后的效果图

3. 绘制圆形

在 canvas 中绘制圆形，将使用以下方法。

arc(x,y,r,start,stop)

实际上在绘制圆形时使用了"ink"的方法，比如 stroke() 或者 fill()。

【案例 2-9】使用 arc() 绘制 Canvas 圆形

极客学院在线视频学习网址：
http://www.jikexueyuan.com/course/1299_1.html
手机扫描二维码

HTML5 绘制圆形

代码清单 2-9

```
<script>
var c=document.getElementById("myCanvas");
var ctx=c.getContext("2d");
ctx.beginPath();
ctx.arc(95,50,40,0,2*Math.PI);
ctx.stroke();
</script>
```

✧ 代码解析

arc() 方法创建弧/曲线（用于创建圆或部分圆）。

如图 2-10 所示，如需通过 arc() 来创建圆，请把起始角设置为 0，结束角设置为 2*Math.PI。使用 stroke() 或 fill() 方法在画布上绘制实际的弧，对应的语法格式如表 2-7 所示，参数如表 2-8 所示。

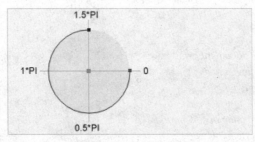

图 2-10 arc() 角度图

中心：
arc(100,75,50,0*Math.PI,1.5*Math.PI)
起始角：
arc(100,75,50,0,1.5*Math.PI)
结束角：
arc(100,75,50,0*Math.PI,1.5*Math.PI)。

表 2-7　arc()语法

JavaScript 语法	context.arc(x, y, r, sAngle, eAngle, counterclockwise);

表 2-8　arc()参数

参数	描述
x	圆的中心的 x 坐标
y	圆的中心的 y 坐标
r	圆的半径
sAngle	起始角，以弧度计（弧的圆形的三点钟位置是 0 度）
eAngle	结束角，以弧度计
counterclockwise	可选。规定应该逆时针还是顺时针绘图。False = 顺时针，true = 逆时针

本案例运行的结果如图 2-11 所示。

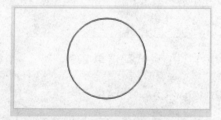

图 2-11　代码清单 2-9 运行后的效果图

【案例 2-10】使用 arc 函数绘制复杂 Canvas 圆形

代码清单 2-10

```
<!DOCTYPE html>
<html lang="en">
<head>
<meta charset="UTF-8">
<title></title>
<script>
        function draw(id){
            var canvas = document.getElementById(id);
            if(canvas == null){
                return false;
            }
            var context = canvas.getContext("2d");
            context.fillStyle = "#eeeeef";
            context.fillRect(0,0,600,700);
```

```
            for(var i =0; i<=10; i++){//循环不断绘制圆形
context.beginPath();
context.arc(i*25,i*25,i*10,0,Math.PI*2,true);
context.closePath();
                context.fillStyle = "rgba(255,0,0,0.25)";
context.fill();
            }
        }
</script>
</head>
<body onload="draw('canvas')">
<canvas id="canvas" width="300" height="400"></canvas>
</body>
</html>
```

❖ 代码解析

本案例主要说明 HTML5 绘制圆形的过程，主要包含四个步骤，开始创建路径；创建图形的路径；路径创建完成后，关闭路径和设定绘制样式；调用绘制方法，绘制路径。代码的解析请查看源代码注释，程序运行后的效果如图 2-12 所示。

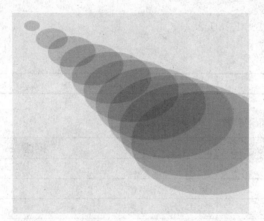

图 2-12　代码清单 2-10 运行后的效果图

4．绘制贝塞尔曲线

【案例 2-11】用 quadraticCurveTo 绘制 Canvas 贝塞尔曲线

代码清单 2-11

```
<script>
var c=document.getElementById("myCanvas");
var ctx=c.getContext("2d");
ctx.beginPath();
ctx.moveTo(20,20);
ctx.quadraticCurveTo(20,100,200,20);
ctx.stroke();
</script>
```

❖ 代码解析

quadraticCurveTo()方法通过使用表示二次贝塞尔曲线的指定控制点，向当前路径添加一个点。

如图 2-13 所示，二次贝塞尔曲线需要两个点。第一个点是用于二次贝塞尔计算中的控制点，第二个点是曲线的结束点。曲线的开始点是当前路径中最后一个点。如果路径不存在，那么请使用 beginPath() 和 moveTo() 方法来定义开始点，对应的语法格式如表 2-9 所示，参数如表 2-10 所示。

图 2-13　贝塞尔曲线路径

开始点：
moveTo(20,20)
控制点：
quadraticCurveTo(20,100,200,20)
结束点：
quadraticCurveTo(20,100,200,20)。

表 2-9　quadraticCurveTo() 语法

JavaScript 语法	context.quadraticCurveTo(cpx,cpy,x,y);

表 2-10　quadraticCurveTo() 参数

参数	描述
cpx	贝塞尔控制点的 x 坐标
cpy	贝塞尔控制点的 y 坐标
x	结束点的 x 坐标
y	结束点的 y 坐标

案例运行结果如图 2-14 所示。

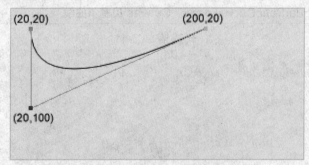

图 2-14　代码清单 2-11 运行后的效果图

【案例 2-12】用 bezierCurveTo( ) 绘制 Canvas 贝塞尔曲线

极客学院

极客学院在线视频学习网址：
http://www.jikexueyuan.com/course/1299_3.html
手机扫描二维码

使用 bezierCurveTo 绘制贝塞尔曲线

代码清单 2-12

```
<script>

var c=document.getElementById("myCanvas");
var ctx=c.getContext("2d");
ctx.beginPath();
ctx.moveTo(20,20);
ctx.bezierCurveTo(20,100,200,100,200,20);
ctx.stroke();


</script>
```

❖ 代码解析

bezierCurveTo() 方法通过使用表示三次贝塞尔曲线的指定控制点，向当前路径添加一个点。

如图 2-15 所示，三次贝塞尔曲线需要三个点。前两个点是用于三次贝塞尔计算中的控制点，第三个点是曲线的结束点。曲线的开始点是当前路径中最后一个点。如果路径不存在，那么请使用 beginPath() 和 moveTo() 方法来定义开始点，对应的语法格式如表 2-11 所示，参数如表 2-12 所示。

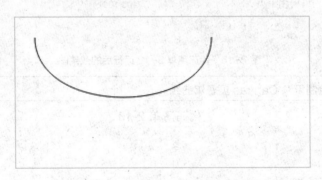

图 2-15　bezierCurveTo 曲线路径图

开始点：
moveTo(20,20)
控制点 1：
bezierCurveTo(20,100,200,100,200,20)
控制点 2：
bezierCurveTo(20,100,200,100,200,20)
结束点：

bezierCurveTo(20,100,200,100,200,20)。

请查看 quadraticCurveTo()方法。它有一个控制点，而不是两个。

表 2-11 bezierCurveTo()语法

| JavaScript 语法 | $context.bezierCurveTo(cp1x, cp1y, cp2x, cp2y, x, y);$ |

表 2-12 bezierCurveTo()参数

参数	描述
$cp1x$	第一个贝塞尔控制点的 x 坐标
$cp1y$	第一个贝塞尔控制点的 y 坐标
$cp2x$	第二个贝塞尔控制点的 x 坐标
$cp2y$	第二个贝塞尔控制点的 y 坐标
x	结束点的 x 坐标
y	结束点的 y 坐标

代码运行结果如图 2-16 所示。

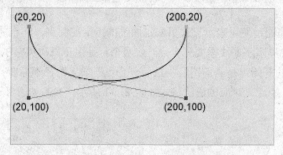

图 2-16 代码清单 2-12 运行后的效果图

【案例 2-13】绘制复杂 Canvas 贝塞尔曲线

代码清单 2-13

```
<!DOCTYPE html>
<html lang="en">
<head>
<meta charset="UTF-8">
<title></title>
<script>
    function draw(id){
        var canvas = document.getElementById(id);
        if(canvas == null){
            return false;
        }
        var context = canvas.getContext("2d");
        context.fillStyle = "#eeeeef";
        context.fillRect(0,0,300,400);
```

```
                var dx = 150;
                var dy = 150;
                var s =100;
                context.beginPath();
                context.fillStyle = "rgb(100,255,100)";
                var x = Math.sin(0);
                var y = Math.cos(0);
                var dig = Math.PI/15*11;
                context.moveTo(dx,dy);
                for(var i=0; i<30; i++){
                    var x = Math.sin(i*dig);
                    var y = Math.cos(i*dig);
context.bezierCurveTo(dx+x*s,dy+y*s-100,dx+x*s+100,dy+y*s,dx+x*s,dy+y*s);
                }
                context.closePath();
                context.fill();
                context.stroke();
        }
</script>
</head>
<body onload="draw('canvas')">
<canvas id="canvas" width="300" height="400"></canvas>
</body>
</html>
```

✧ 代码解析

```
for(var i=0; i<30; i++){
                    var x = Math.sin(i*dig);
                    var y = Math.cos(i*dig);
context.bezierCurveTo(dx+x*s,dy+y*s-100,dx+x*s+100,dy+y*s,dx+x*s,dy+y*s);
                }
```

用循环的方式设置 30 条不同的贝塞尔曲线，代码运行效果如图 2-17 所示。

图 2-17　代码清单 2-13 运行后的效果图

2.2.2　绘制 Canvas 变形图形

渐变可用于填充矩形、圆形、线条、文本等，可以自己定义不同的颜色。以下有两种不同的方式来设置 Canvas 渐变。

createLinearGradient(x,y,x1,y1)：创建线条渐变；

createRadialGradient(x,y,r,x1,y1,r1)：创建一个径向/圆渐变。

当使用渐变对象时，必须使用两种或两种以上的停止颜色。

addColorStop()方法指定颜色停止，参数使用坐标来描述，可以是 0 至 1。

使用渐变，设置 fillStyle 或 strokeStyle 的值为渐变，然后绘制形状，如矩形、文本或一条线。

【案例 2-14】使用 createLinearGradient()绘制渐变效果

极客学院在线视频学习网址：
http://www.jikexueyuan.com/course/1315_1.html
手机扫描二维码

HTML5 Canvas 绘制渐变图形

代码清单 2-14

```
<script>
    var c=document.getElementById("myCanvas");
    var ctx=c.getContext("2d");
    var grd=ctx.createLinearGradient(0,0,170,0);
    grd.addColorStop(0,"black");
    grd.addColorStop(0.5,"red");
    grd.addColorStop(1,"white");
    ctx.fillStyle=grd;
    ctx.fillRect(20,20,150,100);
</script>
```

✧ 代码解析

本案例定义一个从黑到红再到白的渐变，作为矩形的填充样式。使用 createLinearGradient()方法创建线性的渐变对象，对应的语法格式如表 2-13 所示，参数如表 2-14 所示。

 请使用该对象作为 strokeStyle 或 fillStyle 属性的值；请使用 addColorStop()方法规定不同的颜色，以及在 gradient 对象中的何处定位颜色。

表 2-13　createLinearGradient()语法

JavaScript 语法	context.createLinearGradient(x0,y0,x1,y1);

表 2-14　createLinearGradient()参数

参数	描述
x0	渐变开始点的 x 坐标
y0	渐变开始点的 y 坐标
x1	渐变结束点的 x 坐标
y1	渐变结束点的 y 坐标

案例运行结果如图 2-18 所示。

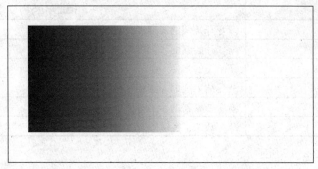

图 2-18　代码清单 2-14 运行后的效果图

【案例 2-15】使用 createRadialGradient() 绘制 Canvas 渐变效果

使用渐变填充矩形，创建一个径向/圆渐变。

极客学院在线视频学习网址：
http://www.jikexueyuan.com/course/1315_2.html
手机扫描二维码

HTML5 Canvas 绘制径向渐变

代码清单 2-15

```
<script>
    var c=document.getElementById("myCanvas");
    var ctx=c.getContext("2d");
    var grd=ctx.createRadialGradient(75,50,5,90,60,100);
    grd.addColorStop(0,"red");
    grd.addColorStop(1,"white");
    ctx.fillStyle=grd;
    ctx.fillRect(10,10,150,100);
</script>
```

◆ 代码解析

createRadialGradient() 方法创建放射状/圆形渐变对象，渐变可用于填充矩形、圆形、线条、文本等，对应的语法格式如表 2-15 所示，参数如表 2-16 所示。

 请使用该对象作为 strokeStyle 或 fillStyle 属性的值，使用 addColorStop() 方法规定不同的颜色，以及在 gradient 对象中的何处定位颜色。

表 2-15　createRadialGradient() 语法

JavaScript 语法	*context*.createRadialGradient(*x0, y0, r0, x1, y1, r1*);

表 2-16 createRadialGradient()参数

参数	描述
x0	渐变开始圆的 x 坐标
y0	渐变开始圆的 y 坐标
r0	开始圆的半径
x1	渐变结束圆的 x 坐标
y1	渐变结束圆的 y 坐标
r1	结束圆的半径

案例运行结果如图 2-19 所示。

图 2-19 代码清单 2-15 运行后的效果图

【案例 2-16】绘制复杂渐变效果

代码清单 2-16

```
<script>
    function draw(id){
        var canvas = document.getElementById(id);
        var context = canvas.getContext("2d");
        var g1 = context.createLinearGradient(0,0,0,300);
        g1.addColorStop(0,"rgb(255,255,0)");
        g1.addColorStop(1,"rgb(0,255,255)");
        context.fillStyle = g1;//背景矩形的渐变效果
        context.fillRect(0,0,500,500);
        var g2 = context.createLinearGradient(0,0,300,0);
        g2.addColorStop(0,"rgba(0,0,255,0.5)");
        g2.addColorStop(1,"rgba(255,0,0,0.5)");
        for(var i = 1; i<10; i++){//使用for循环生成多个渐变图形
            context.beginPath();
            context.fillStyle = g2;
            context.arc(i*25,i*25,i*10,0,Math.PI*2,true);
            context.closePath();
            context.fill();
        }
    }
</script>
```

✧ 代码解析

本案例主要使用 for(var i = 1; i<10; i++)来绘制 9 个径向渐变效果图，其中图 2-20（a）背景效果

使用 context.fillStyle = g1，圆形渐变效果使用 context.fillStyle = g2。为了清楚地查看圆形图形的渐变效果，可以注释代码段：

//var g1 = context.createLinearGradient(0,0,0,300);

//g1.addColorStop(0,"rgb(255,255,0)");

//g1.addColorStop(1,"rgb(0,255,255)");

//context.fillStyle = g1;//背景矩形的渐变效果

案例运行后的效果如图 2-20（b）所示。

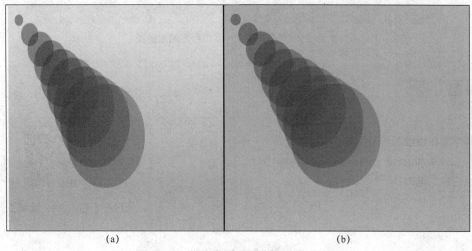

图 2-20　代码清单 2-16 运行后的效果图

【案例 2-17】使用单独的函数来绘制渐变效果

代码清单 2-17

```
<script>
    function draw(id){
        var canvas = document.getElementById(id);
        if(canvas == null){
            return false;
        }
        var context = canvas.getContext("2d");
        var g1 = context.createRadialGradient(400,50,50,400,50,400);
        g1.addColorStop(0.1,"rgb(255,255,0)");
        g1.addColorStop(0.3,"rgb(255,0,255)");
        g1.addColorStop(1,"rgb(0,255,255)");
        context.fillStyle=g1;
        context.fillRect(0,0,500,500);
    }
</script>
```

◆　代码解析

本案例使用函数的形式来设计渐变效果，包装好的函数可以在任何位置调用，案例的运行效果如图 2-21 所示。

【案例 2-18】使用 scale() 绘制变形图形

图 2-21 代码清单 2-17 运行后的效果图

极客学院在线视频学习网址：
http://www.jikexueyuan.com/course/1315_3.html
手机扫描二维码

HTML5 Canvas 绘制变形图形

代码清单 2-18

```
<script>
    var c=document.getElementById("myCanvas");
    var ctx=c.getContext("2d");
    ctx.strokeRect(5,5,25,15);
    ctx.scale(2,2);
    ctx.strokeRect(5,5,25,15);
</script>
```

◆ 代码解析

scale()方法缩放当前绘图至更大或更小，对应的语法格式如表 2-17 所示，参数如表 2-18 所示。

如果对绘图进行缩放，所有之后的绘图也会被缩放。定位也会被缩放。如果使用 scale(2,2)，那么绘图将定位于距离画布左上角两倍远的位置。

表 2-17　scale()语法

JavaScript 语法	*context*.scale(*scalewidth*, *scaleheight*);

表 2-18　scale()参数

参数	描述
scalewidth	缩放当前绘图的宽度（1=100%，0.5=50%，2=200%，依次类推
scaleheight	缩放当前绘图的高度（1=100%，0.5=50%，2=200%，依次类推

案例的运行效果如图 2-22 所示。

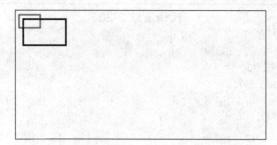

图 2-22　代码清单 2-18 运行后的效果图

【案例 2-19】使用 rotate() 绘制变形图形

代码清单 2-19

```
<script>
    var c=document.getElementById("myCanvas");
    var ctx=c.getContext("2d");
    ctx.rotate(20*Math.PI/180);
    ctx.fillRect(50,20,100,50);
</script>
```

❖ 代码解析

rotate() 方法旋转当前的绘图，对应的语法格式如表 2-19 所示，参数如表 2-20 所示。

 旋转只会影响到旋转完成后的绘图。

表 2-19　rotate() 语法

JavaScript 语法	*context*.rotate(*angle*);

表 2-20　rotate() 参数

参数	描述
angle	旋转角度，以弧度计 如需将角度转换为弧度，请使用 *degrees**Math.PI/180 公式进行计算 实例：如需旋转 5 度，可规定下面的公式：5*Math.PI/180

案例的运行效果如图 2-23 所示。

图 2-23　代码清单 2-19 运行后的效果图

【案例 2-20】使用 translate()绘制变形图形

代码清单 2-20

```
<script>
    var c=document.getElementById("myCanvas");
    var ctx=c.getContext("2d");
    ctx.fillRect(10,10,100,50);
    ctx.translate(70,70);
    ctx.fillRect(10,10,100,50);
</script>
```

✧ 代码解析

translate()方法重新映射画布上的(0,0)位置。当在 translate()之后调用诸如 fillRect()之类的方法时，值会添加到 x 和 y 坐标值上，对应的语法格式如表 2-21 所示，参数如表 2-22 所示。

表 2-21　translate()语法

JavaScript 语法	$context$.translate(x, y);

表 2-22　translate()参数

参数	描述
x	添加到水平坐标 x 上的值
y	添加到垂直坐标 y 上的值

案例的运行效果如图 2-24 所示。

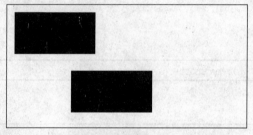

图 2-24　代码清单 2-20 运行后的效果图

【案例 2-21】使用 transform()绘制变形图形

绘制一个矩形，通过 transform() 添加一个新的变换矩阵，再次绘制矩形，添加一个新的变换矩阵，然后再次绘制矩形。请注意，每当调用 transform() 时，它都会在前一个变换矩阵上构建。

代码清单 2-21

```
<script>
    var c=document.getElementById("myCanvas");
    var ctx=c.getContext("2d");
    ctx.fillStyle="yellow";
    ctx.fillRect(0,0,250,100);
    ctx.transform(1,0.5,-0.5,1,30,10);
    ctx.fillStyle="red";
    ctx.fillRect(0,0,250,100);
    ctx.transform(1,0.5,-0.5,1,30,10);
    ctx.fillStyle="blue";
```

```
ctx.fillRect(0,0,250,100);
</script>
```

案例的运行效果如图 2-25 所示。

图 2-25　代码清单 2-21 运行后的效果图

◆ 代码解析

画布上的每个对象都拥有一个当前的变换矩阵，transform() 方法替换当前的变换矩阵。它以下面描述的矩阵来操作当前的变换矩阵：

a	c	e
b	d	f
0	0	1

换句话说，transform()允许缩放、旋转、移动并倾斜当前的环境。

该变换只会影响 transform()方法调用之后的绘图。transform()方法的行为相对于由 rotate()、scale()、translate()或 transform()完成的其他变换。例如：如果已经将绘图设置为放到两倍，则 transform()方法会把绘图放大两倍，最终将放大四倍，对应的语法格式如表 2-23 所示，参数如表 2-24 所示。

请查看 setTransform()方法，它不会相对于其他变换来发生行为。

表 2-23　setTransform ()语法

JavaScript 语法	*context*.transform(*a,b,c,d,e,f*);

表 2-24　setTransform ()参数

参数	描述
a	水平缩放绘图
b	水平倾斜绘图
c	垂直倾斜绘图
d	垂直缩放绘图
e	水平移动绘图
f	垂直移动绘图

【案例 2-22】绘制复杂变形图形

代码清单 2-22

```
<script>
    function draw(id){
        var canvas = document.getElementById(id);
        if(canvas == null){
            return false;
        }
        var context = canvas.getContext("2d");
        context.fillStyle = "#eeeeef";
        context.fillRect(0,0,500,500);
        context.translate(200,50);
        context.fillStyle = "rgba(255,0,0,0.25)";
        for(var i =0; i<50; i++){
            context.translate(25,25);
            context.scale(0.95,0.95);
            context.rotate(Math.PI/10);
            context.fillRect(0,0,100,50);
        }
    }
</script>
```

❖ 代码解析

本案例主要使用循环语句 for(var i =0; i<50; i++)绘制 50 次不同的变形图形,案例的运行效果如图 2-26 所示。

图 2-26　代码清单 2-22 运行后的效果图

2.2.3　处理 Canvas 图形

【案例 2-23】使用 globalCompositeOperation 参数对图形进行处理

代码清单 2-23

```
<script>
    var gco=new Array();
    gco.push("source-atop");
    gco.push("source-in");
    gco.push("source-out");
    gco.push("source-over");
```

```
            gco.push("destination-atop");
            gco.push("destination-in");
            gco.push("destination-out");
            gco.push("destination-over");
            gco.push("lighter");
            gco.push("copy");
            gco.push("xor");
            for (n=0;n<gco.length;n++)
            {
                document.write("<div id='p_" + n + "' style='float:left;'>" + gco[n] + ":<br>");
                var c=document.createElement("canvas");
                c.width=120;
                c.height=100;
                document.getElementById("p_" + n).appendChild(c);
                var ctx=c.getContext("2d");
                ctx.fillStyle="blue";
                ctx.fillRect(10,10,50,50);
                ctx.globalCompositeOperation=gco[n];
                ctx.beginPath();
                ctx.fillStyle="red";
                ctx.arc(50,50,30,0,2*Math.PI);
                ctx.fill();
                document.write("</div>");
            }
        </script>
```

❖ 代码解析

globalCompositeOperation 属性设置或返回如何将一个源（新的）图像绘制到目标（已有的）的图像上，对应的语法格式如表 2-25 所示，属性如表 2-26 所示。

源图像=打算放置到画布上的绘图。

目标图像=已经放置在画布上的绘图。

表 2-25　globalCompositeOperation 语法

默认值	source-over
JavaScript 语法	*context*.globalCompositeOperation="source-in";

表 2-26　globalCompositeOperation 属性值

值	描述
source-over	默认。在目标图像上显示源图像
source-atop	在目标图像顶部显示源图像。源图像位于目标图像之外的部分是不可见的
source-in	在目标图像中显示源图像。只有目标图像之内的源图像部分会显示，目标图像是透明的
source-out	在目标图像之外显示源图像。只有目标图像之外的源图像部分会显示，目标图像是透明的
destination-over	在源图像上显示目标图像
destination-atop	在源图像顶部显示目标图像。目标图像位于源图像之外的部分是不可见的
destination-in	在源图像中显示目标图像。只有源图像之内的目标图像部分会被显示，源图像是透明的

续表

值	描述
destination-out	在源图像之外显示目标图像。只有源图像之外的目标图像部分会被显示，源图像是透明的
lighter	显示源图像 + 目标图像
copy	显示源图像。忽略目标图像
xor	使用异或操作对源图像与目标图像进行组合

案例的运行效果如图 2-27 所示。

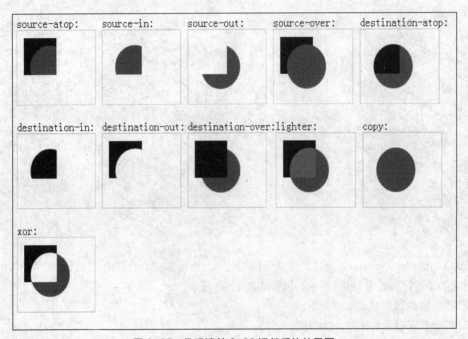

图 2-27 代码清单 2-23 运行后的效果图

【案例 2-24】使用 shadowColor 参数对图形进行处理

代码清单 2-24

```
<script>
    var c=document.getElementById("myCanvas");
    var ctx=c.getContext("2d");
    ctx.shadowBlur=10;
    ctx.shadowOffsetX=20;
    ctx.shadowColor="black";
    ctx.fillStyle="red";
    ctx.fillRect(20,20,100,80);
</script>
```

◆ 代码解析

shadowColor 属性设置或返回用于阴影的颜色，对应的语法格式如表 2-27 所示，参数如表 2-28 所示。

请将 shadowColor 属性与 shadowBlur 属性一起使用，来创建阴影。

请通过使用 shadowOffsetX 和 shadowOffsetY 属性来调节阴影效果。

表 2-27　shadowColor 语法

默认值	#000000
JavaScript 语法	*context*.shadowColor=*color*;

表 2-28　shadowColor 属性值

值	描述
color	用于阴影的 CSS 颜色值。默认值是#000000

shadowBlur 属性设置或返回阴影的模糊级数，对应的语法格式如表 2-29 所示，参数如表 2-30 所示。

表 2-29　shadowBlur 语法

默认值	0
JavaScript 语法	*context*.shadowBlur=*number*;

表 2-30　shadowBlur 属性值

值	描述
number	阴影的模糊级数

shadowOffsetX 属性设置或返回阴影与形状的水平距离，对应的语法格式如表 2-31 所示，参数如表 2-32 所示。

shadowOffsetX=0 指示阴影位于形状的正下方。
shadowOffsetX=20 指示阴影位于形状 left 位置右侧的 20 像素处。
shadowOffsetX=-20 指示阴影位于形状 left 位置左侧的 20 像素处。

 如需调整阴影与形状的垂直距离，请使用 shadowOffsetY 属性。

表 2-31　shadowOffsetX 语法

默认值	0
JavaScript 语法	*context*.shadowOffsetX=*number*;

表 2-32　shadowOffsetX 属性值

值	描述
number	正值或负值，定义阴影与形状的水平距离

案例的运行效果如图 2-28 所示。

图 2-28 代码清单 2-24 运行后的效果图

【案例 2-25】用单独函数绘制五角星阴影

代码清单 2-25

```
<script>
    function draw(id){
        var canvas = document.getElementById(id);
        var context = canvas.getContext("2d");
        context.fillStyle = "#eeeeef";
        context.fillRect(0,0,500,500);
        context.shadowOffsetX = 20;
        context.shadowOffsetY = 10;
        context.shadowColor = "rgba(100,100,100,0.5)";
        context.shadowBlur = 3.5;
        context.translate(0,50);
        for(var i =0; i<3; i++){
            context.translate(100,100);
            create5Start(context);
            context.fill();
        }
    }

    function create5Start(context){
        var dx = 100;
        var dy = 0;
        var s = 50;
        context.beginPath();
        context.fillStyle = "rgba(255,0,0,0.5)";
        var dig = Math.PI/5*4;
        for(var i = 0; i<5; i++){
            var x = Math.sin(i*dig);
            var y = Math.cos(i*dig);
            context.lineTo(dx+x*s,dy+y*s);
        }
        context.closePath();
    }
</script>
```

◆ 代码解析

本案例结合变形操作、阴影效果属性来画图。

首先在 draw(id)函数中设置所有绘图阴影效果。

```
context.shadowOffsetX = 20;
context.shadowOffsetY = 10;
context.shadowColor = "rgba(100,100,100,0.5)";
context.shadowBlur = 3.5;
```

然后使用

```
for(var i =0; i<3; i++){
    context.translate(100,100);
    create5Start(context);
    context.fill();
}
```

循环绘制 3 个五角星形图形，其中画五角星函数包装在 create5Start(context)函数中：

```
for(var i = 0; i<5; i++){
    var x = Math.sin(i*dig);
    var y = Math.cos(i*dig);
    context.lineTo(dx+x*s,dy+y*s);
}
```

该函数使用循环用画线的方式画出五角星形图形，运行效果如图 2-29 所示。

图 2-29　代码清单 2-25 运行后的效果图

2.3　Canvas 文本

2.3.1　绘制文字

【案例 2-26】使用 fillText()函数绘制文字

代码清单 2-26

```
<script>
    var c=document.getElementById("myCanvas");
    var ctx=c.getContext("2d");
    ctx.font="20px Georgia";
    ctx.fillText("Hello World!",10,50);
    ctx.font="30px Verdana";
    // Create gradient
    var gradient=ctx.createLinearGradient(0,0,c.width,0);
    gradient.addColorStop("0","magenta");
```

```
        gradient.addColorStop("0.5","blue");
        gradient.addColorStop("1.0","red");
        // Fill with gradient
        ctx.fillStyle=gradient;
        ctx.fillText("Big smile!",10,90);
    </script>
```

◆ 代码解析

fillText() 方法在画布上绘制填色的文本。文本的默认颜色是黑色,对应的语法格式如表 2-33 所示,参数如表 2-34 所示。

请使用 font 属性来定义字体和字号,并使用 fillStyle 属性以另一种颜色/渐变来渲染文本。

表 2-33　fillText()语法

JavaScript 语法	$context.\text{fillText}(text, x, y, maxWidth)$;

表 2-34　fillText()参数

参数	描述
$text$	规定在画布上输出的文本.
x	开始绘制文本的 x 坐标位置(相对于画布)
y	开始绘制文本的 y 坐标位置(相对于画布)
$maxWidth$	可选。允许的最大文本宽度,以像素计

案例运行效果如图 2-30 所示。

Hello World!
Big smile!

图 2-30　代码清单 2-26 运行后的效果图

【案例 2-27】使用 strokeText()函数绘制文字

代码清单 2-27

```
    <script>
        var c=document.getElementById("myCanvas");
        var ctx=c.getContext("2d");
        ctx.font="20px Georgia";
        ctx.strokeText("Hello World!",10,50);
        ctx.font="30px Verdana";
```

```
    // Create gradient
    var gradient=ctx.createLinearGradient(0,0,c.width,0);
    gradient.addColorStop("0","magenta");
    gradient.addColorStop("0.5","blue");
    gradient.addColorStop("1.0","red");
    // Fill with gradient
    ctx.strokeStyle=gradient;
    ctx.strokeText("Big smile!",10,90);
</script>
```

- ◆ 代码解析

strokeText() 方法在画布上绘制文本（无填充色）。文本的默认颜色是黑色，对应的语法格式如表 2-35 所示，参数如表 2-36 所示。

 请使用 font 属性来定义字体和字号，并使用 strokeStyle 属性以另一种颜色/渐变来渲染文本。

表 2-35　strokeText ()语法

JavaScript 语法	$context.\text{strokeText}(text, x, y, maxWidth);$

表 2-36　strokeText ()参数

参数	描述
$text$	规定在画布上输出的文本
x	开始绘制文本的 x 坐标位置（相对于画布）
y	开始绘制文本的 y 坐标位置（相对于画布）
$maxWidth$	可选。允许的最大文本宽度，以像素计

案例运行效果如图 2-31 所示。

图 2-31　代码清单 2-27 运行后的效果图

2.3.2　设置文字格式

【案例 2-28】文字大小设置

代码清单 2-28

```
</canvas>
<script type="text/javascript">
var c=document.getElementById("myCanvas");
var ctx=c.getContext("2d");

ctx.beginPath();
//设定文字大小为30px
ctx.font="30px Arial";
ctx.fillText("Hello World",100,50);

ctx.beginPath();
//设定文字大小为50px
ctx.font="50px Arial";
ctx.fillText("Hello World",100,150);

ctx.beginPath();
//设定文字大小为100px
ctx.font="70px Arial";
ctx.fillText("Hello World",100,250);
</script>
```

❖ 代码解析

font 属性设置或返回画布上文本内容的当前字体属性。

font 属性使用的语法与 CSS font 属性相同，对应的语法格式如表 2-37 所示，参数如表 2-38 所示。

表 2-37 font 语法

默认值	10px sans-serif
JavaScript 语法	*context*.font="italic small-caps bold 12px arial";

表 2-38 font 属性值

值	描述
font-style	规定字体样式。可能的值：normal、italic、oblique
font-variant	规定字体变体。可能的值：normal、small-caps
font-weight	规定字体的粗细。可能的值：normal、bold、bolder、lighter、100、200、300、400、500、600、700、800、900
font-size/line-height	规定字号和行高，以像素计
font-family	规定字体系列
caption	使用标题控件的字体（比如按钮、下拉列表等）
icon	使用用于标记图标的字体
menu	使用用于菜单中的字体（下拉列表和菜单列表）
message-box	使用用于对话框中的字体
small-caption	使用用于标记小型控件的字体
status-bar	使用用于窗口状态栏中的字体

案例运行效果如图 2-32 所示。

图 2-32 代码清单 2-28 运行后的效果图

【案例 2-29】文字字体设置

代码清单 2-29

```
<script type="text/javascript">
   var c=document.getElementById("myCanvas");
   var ctx=c.getContext("2d");

   ctx.beginPath();
   //设定文字字体为Arial
   ctx.font="30px Arial";
   ctx.fillText("Hello World (Arial)",50,50);

   ctx.beginPath();
   //设定文字字体为Verdana
   ctx.font="30px Verdana";
   ctx.fillText("Hello World (Verdana)",50,100);

   ctx.beginPath();
   //设定文字字体为Times New Roman
   ctx.font="30px Times New Roman";
   ctx.fillText("Hello World (Times New Roman)",50,150);

   ctx.beginPath();
   //设定文字字体为Courier New
   ctx.font="30px Courier New";
   ctx.fillText("Hello World (Courier New)",50,200);

   ctx.beginPath();
   //设定文字字体为serif
   ctx.font="30px serif";
   ctx.fillText("Hello World (serif)",50,250);

   ctx.beginPath();
   //设定文字字体为sans-serif
   ctx.font="30px sans-serif";
   ctx.fillText("Hello World (sans-serif)",50,300);
</script>
```

✧ 代码解析

Font 用法与属性值查看【案例 2-28】，本案例运行效果如图 2-33 所示。

Hello World (Arial)
Hello World (Verdana)
Hello World (Times New Roman)
Hello World (Courier New)
Hello World (serif)
Hello World (sans-serif)

图 2-33　代码清单 2-29 运行后的效果图

【案例 2-30】文字粗体设置

代码清单 2-30

```
<script type="text/javascript">
var c=document.getElementById("myCanvas");
var ctx=c.getContext("2d");

ctx.beginPath();
//设定font-weight为normal
ctx.font='normal 30px Arial';
ctx.fillText("Hello World (normal)",50,50);

ctx.beginPath();
//设定font-weight为bold
ctx.font='bold 30px Arial';
ctx.fillText("Hello World (bold)",50,90);

ctx.beginPath();
//设定font-weight为bolder
ctx.font='bolder 30px Arial';
ctx.fillText("Hello World (bolder)",50,130);

ctx.beginPath();
//设定font-weight为lighter
ctx.font='lighter 30px Arial';
ctx.fillText("Hello World (lighter)",50,170);

ctx.beginPath();
//设定font-weight为100
ctx.font='100 30px Arial';
ctx.fillText("Hello World (100)",50,210);

ctx.beginPath();
//设定font-weight为600
```

```
ctx.font='600 30px Arial';
ctx.fillText("Hello World (600)",50,250);

ctx.beginPath();
//设定font-weight为900
ctx.font='900 30px Arial';
ctx.fillText("Hello World (900)",50,290);
</script>
```

❖ 代码解析

Font 用法与属性值查看【案例 2-28】，本案例运行效果如图 2-34 所示。

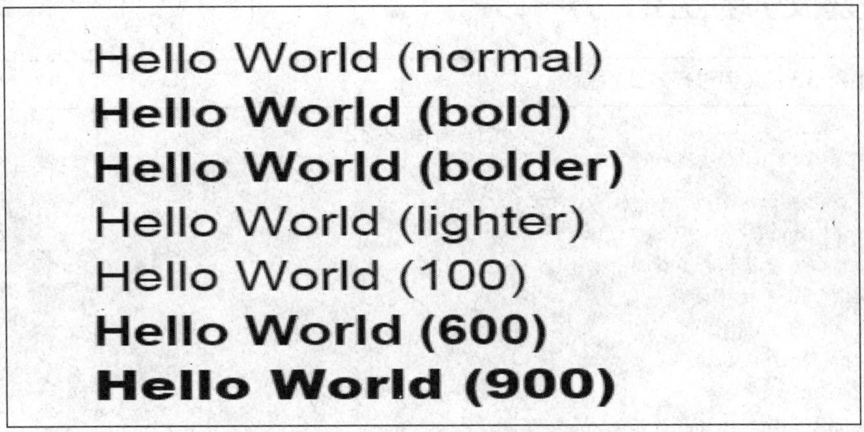

Hello World (normal)
Hello World (bold)
Hello World (bolder)
Hello World (lighter)
Hello World (100)
Hello World (600)
Hello World (900)

图 2-34　代码清单 2-30 运行后的效果图

【案例 2-31】文字斜体设置

代码清单 2-31

```
<script type="text/javascript">
var c=document.getElementById("myCanvas");
var ctx=c.getContext("2d");

ctx.beginPath();
//设定font-weight为normal
ctx.font='normal 30px Arial';
ctx.fillText("Hello World (normal)",50,50);

ctx.beginPath();
//设定font-style为italic
ctx.font='italic 30px Arial';
ctx.fillText("Hello World (italic)",50,90);

ctx.beginPath();
//设定font-style为oblique
ctx.font='oblique 30px Arial';
ctx.fillText("Hello World (oblique)",50,130);
</script>
```

❖ 代码解析

Font 用法与属性值查看【案例 2-28】，本案例运行效果如图 2-35 所示。

> Hello World (normal)
> *Hello World (italic)*
> *Hello World (oblique)*

图 2-35 代码清单 2-31 运行后的效果图

2.3.3 设置文字对齐方式

【案例 2-32】文字对齐方式设置

代码清单 2-32

```
<script>
    var c=document.getElementById("myCanvas");
    var ctx=c.getContext("2d");
    // Create a red line in position 150
    ctx.strokeStyle="red";
    ctx.moveTo(150,20);
    ctx.lineTo(150,170);
    ctx.stroke();
    ctx.font="15px Arial";
    // Show the different textAlign values
    ctx.textAlign="start";
    ctx.fillText("textAlign=start",150,60);
    ctx.textAlign="end";
    ctx.fillText("textAlign=end",150,80);
    ctx.textAlign="left";
    ctx.fillText("textAlign=left",150,100);
    ctx.textAlign="center";
    ctx.fillText("textAlign=center",150,120);
    ctx.textAlign="right";
    ctx.fillText("textAlign=right",150,140);
</script>
```

◆ 代码解析

textAlign 属性根据锚点，设置或返回文本内容的当前对齐方式。

通常，文本会从指定位置开始，不过，如果设置为 textAlign="right" 并将文本放置到位置 150，那么会在位置 150 结束，对应的语法格式如表 2-39 所示，参数如表 2-40 所示。

 请使用 fillText() 或 strokeText() 方法在画布上实际地绘制并定位文本。

表 2-39 textAlign 语法

默认值	start
JavaScript 语法	*context*.textAlign="center|end|left|right|start";

表 2-40　textAlign 属性值

值	描述
start	默认。文本在指定的位置开始
end	文本在指定的位置结束
center	文本的中心被放置在指定的位置
left	文本在指定的位置开始
right	文本在指定的位置结束

本案例主要是在文本间多画出竖线以突出 textAlign 的显示效果，运行结果如图 2-36 所示。

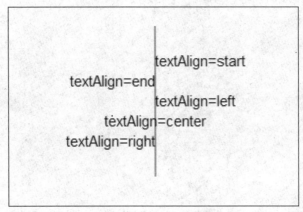

图 2-36　代码清单 2-32 运行后的效果图

【案例 2-33】文本基线设置

代码清单 2-33

```
<script type="text/javascript">
var c=document.getElementById('myCanvas');
var ctx=c.getContext('2d');

ctx.textBaseline='alphabetic';
ctx.font='30px Arial';
ctx.fillText('Hello World',50,50);
ctx.moveTo(0,50);
ctx.lineTo(250,50);
ctx.stroke();

ctx.textBaseline='bottom';
ctx.font='30px Arial';
ctx.fillText('Hello World',50,100);
ctx.moveTo(0,100);
ctx.lineTo(250,100);
ctx.stroke();

ctx.textBaseline='hanging';
ctx.font='30px Arial';
ctx.fillText('Hello World',50,150);
```

```
ctx.moveTo(0,150);
ctx.lineTo(250,150);
ctx.stroke();

ctx.textBaseline='ideographic';
ctx.font='30px Arial';
ctx.fillText('Hello World',50,200);
ctx.moveTo(0,200);
ctx.lineTo(250,200);
ctx.stroke();

ctx.textBaseline='middle';
ctx.font='30px Arial';
ctx.fillText('Hello World',50,250);
ctx.moveTo(0,250);
ctx.lineTo(250,250);
ctx.stroke();

ctx.textBaseline='top';
ctx.font='30px Arial';
ctx.fillText('Hello World',50,300);
ctx.moveTo(0,300);
ctx.lineTo(250,300);
ctx.stroke();
</script>
```

❖ 代码解析

textBaseline 属性设置或返回在绘制文本时的当前文本基线，对应的语法格式如表 2-41 所示，参数如表 2-42 所示。

fillText()和 strokeText()方法在画布上定位文本时，将使用指定的 textBaseline 值。

表 2-41 textBaseline 语法

默认值	alphabetic
JavaScript 语法	*context*.textBaseline="alphabetic|top|hanging|middle|ideographic|bottom";

表 2-42 textBaseline 属性值

值	描述
alphabetic	默认。文本基线是普通的字母基线
top	文本基线是 em 方框的顶端
hanging	文本基线是悬挂基线
middle	文本基线是 em 方框的正中
ideographic	文本基线是表意基线
bottom	文本基线是 em 方框的底端

本案例主要是在文本间多画出横线以突出 textBaseline 的显示效果，案例运行结果如图 2-37 所示。

图 2-37　代码清单 2-33 运行后的效果图

2.4　Canvas 图片

无论开发的是应用程序还是游戏软件，都离不开图片，没有图片就无法让整个页面漂亮起来。开发 HTML5 游戏的时候，游戏中的地图、背景、人物、物品等都是由图片组成的，所以图片的显示和操作都非常重要。本节主要使用 Canvas drawImage 函数、getImageData 和 putImageData 绘制图片，使用 createImageData 新建图片像素。

极客学院在线视频学习网址：
http://www.jikexueyuan.com/course/221_3.html
手机扫描二维码

HTML5Canvas 标签-绘制图片

2.4.1　绘制 drawImage 图片

下面主要用案例来说明 DrawImage 的使用过程。

【案例 2-34】使用 drawImage()函数在画布上绘制图片

代码清单 2-34

```
<body>
img标签：<br />
<imgsrc="images/image.jpg"></img>
<br />Canvas画板：<br />
<canvas id="myCanvas" width="400" height="400">
你的浏览器不支持HTML5
</canvas>
<script type="text/javascript">
```

```
var c=document.getElementById("myCanvas");
varctx=c.getContext("2d");
var image = new Image();
image.src = "images/image.jpg";
image.onload = function(){
ctx.drawImage(image,10,10);
ctx.drawImage(image,110,10,110,110);
ctx.drawImage(image,10,10,50,50,210,10,150,150);
};
</script>
</body>
```

❖ 代码解析

drawImage()函数在 Canvas 画布上绘制图像、画布或视频，也能够绘制图像的某些部分，以及增加或减少图像的尺寸，对应的语法格式如表 2-43～表 2-45 所示，参数如表 2-46 所示。

在画布上定位图像。

表 2-43 drawImage()语法 1

JavaScript 语法	$context$.drawImage(img, x, y);

在画布上定位图像，并规定图像的宽度和高度：

表 2-44 drawImage()语法 2

JavaScript 语法	$context$.drawImage($img, x, y, width, height$);

剪切图像，并在画布上定位被剪切的部分：

表 2-45 drawImage()语法 3

JavaScript 语法	$context$.drawImage($img, sx, sy, swidth, sheight, x, y, width, height$);

表 2-46 说明各参数的使用方法。

表 2-46 drawImage()参数

参数	描述
img	规定要使用的图像、画布或视频
sx	可选。开始剪切的 x 坐标位置
sy	可选。开始剪切的 y 坐标位置
$swidth$	可选。被剪切图像的宽度
$sheight$	可选。被剪切图像的高度
x	在画布上放置图像的 x 坐标位置
y	在画布上放置图像的 y 坐标位置
$width$	可选。要使用的图像的宽度（伸展或缩小图像）
$height$	可选。要使用的图像的高度（伸展或缩小图像）

`<imgsrc="images/image.jpg"></img>`

表示运行 image.jpg 原图用于 Canvas drawIamge()函数画图效果的对比；

`ctx.drawImage(image,10,10);`

表示从坐标(10,10)的位置绘制 image.jpg 图片；

ctx.drawImage(image,110,10,110,110);

表示从坐标(110,10)位置绘制整张 image.jpg 图片到长 110、款 110 的矩形区域内，所以本例的运行效果会有一定的拉升感；

ctx.drawImage(image,10,10,50,50,210,10,150,150);

表示将 image.jpg 图片从(10,10)坐标位置截取(50,50)的宽度和高度，然后将截取到的图片从坐标(210,10)位置开始绘制，放到长 110、宽 110 的矩形区域内。

运行效果如图 2-38 所示。

图 2-38　代码清单 2-34 运行后的效果图

2.4.2　使用 getImageData()和 putImageData()绘制图片

下面主要用案例来说明 getImageData()和 putImageData()函数的使用过程。

【案例 2-35】利用 getImageData()和 putImageData()绘制图片

代码清单 2-35

```
<script type="text/javascript">
var c=document.getElementById("myCanvas");
varctx=c.getContext("2d");
var image = new Image();
image.src = "images/image.jpg";
image,onload = function(){
 ctx.drawImage(image,10,10);
 varimgData=ctx.getImageData(20,20,100,100);
 ctx.putImageData(imgData,10,110);
ctx.putImageData(imgData,90,110,20,20,50,50);
};
</script>
```

◆　代码解析

getImageData() 函数返回 ImageData 对象，该对象拷贝了画布指定矩形的像素数据。ImageData 对象不是图像，它规定了画布上一个部分（矩形），并保存了该矩形内每个像素的信息。对于 ImageData 对象中的每个像素，都存在着四方面的信息，即 RGBA 值：

R：红色（0～255）

G：绿色（0~255）

B：蓝色（0~255）

A：alpha 通道（0~255；0 是透明的，255 是完全可见的）

color/alpha 信息以数组形式存在，并存储于 ImageData 对象的 data 属性中。

getImageData()语法格式如表 2-47~表 2-48 所示。

表 2-47　getImageData()语法

JavaScript 语法	context.getImageData(x, y, width, height);

表 2-48 说明 getImageData()各参数的使用方法。

表 2-48　getImageData()参数

参数	描述
x	开始复制的左上角位置的 x 坐标（以像素计）
y	开始复制的左上角位置的 y 坐标（以像素计）
width	要复制的矩形区域的宽度
height	要复制的矩形区域的高度

在操作完成数组中的 color/alpha 信息之后，可以使用 putImageData() 方法将图像数据（从指定的 ImageData 对象）放回画布上。

putImageData()语法格式如表 2-49 所示。

表 2-49　putImageData ()语法

JavaScript 语法	context.putImageData(imgData, x, y, dirtyX, dirtyY, dirtyWidth, dirtyHeight);

表 2-50 说明 putImageData()各参数的使用方法。

表 2-50　getImageData()参数

参数	描述
imgData	规定要放回画布的 ImageData 对象
x	水平值 x，以像素计，在画布上放置图像的位置
y	垂直值 y，以像素计，在画布上放置图像的位置
dirtyX	可选。ImageData 对象左上角的 x 坐标，以像素计
dirtyY	可选。ImageData 对象左上角的 y 坐标，以像素计
dirtyWidth	可选。ImageData 对象所截取的宽度
dirtyHeight	可选。ImageData 对象所截取的高度

ctx.drawImage(image,10,10);

表示为看出 getImageData()、putImageData()函数绘制图片的不同，用 drawImage()函数从坐标(10,10)的位置绘制 image.jpg 图片；

varimgData=ctx.getImageData(20,20,100,100);

表示使用 getImageData()函数从画板中获取像素数据；

ctx.putImageData(imgData,10,110);

表示将所取得的整个像素数据画到 Canvas 画板以(10,110)为起始坐标的位置上；

ctx.putImageData(imgData,90,110,20,20,100,100);

表示将所取得的像素数据一部分画到画板上：从获取得的像素的(20,20)坐标位置开始截取像素，获取(50,50)长宽的像素区域，然后将截取到的像素画到 Canvas 画板以(90,110)为起始坐标的位置上。

运行效果如图 2-39 所示。

图 2-39　代码清单 2-35 运行后的效果图

2.4.3　使用 createImageData()新建像素

下面主要使用案例来说明 createImageData()函数的使用。

【案例 2-36】使用 createImageData()函数在画布上绘制图片

代码清单 2-36

```javascript
<script type="text/javascript">
var c=document.getElementById("myCanvas");
varctx=c.getContext("2d");
var image = new Image();
image.src = "images/image.jpg";
image.onload = function(){
    ctx.drawImage(image,10,10);
    varimgData=ctx.getImageData(20,20,100,100);

    var imgData01=ctx.createImageData(imgData);
    for (i=0; i<imgData01.width*imgData01.height*4;i+=4){
        imgData01.data[i+0]=255;
        imgData01.data[i+1]=0;
        imgData01.data[i+2]=0;
        imgData01.data[i+3]=255;
    }
    ctx.putImageData(imgData01,10,110);

    var imgData02=ctx.createImageData(100,100);
    for (i=0; i<imgData02.width*imgData02.height*4;i+=4){
        imgData02.data[i+0]=255;
        imgData02.data[i+1]=0;
        imgData02.data[i+2]=0;
        imgData02.data[i+3]=155;
    }
```

```
        ctx.putImageData(imgData02,120,110);
    };
</script>
```

❖ 代码解析

createImageData() 方法创建新的空白 ImageData 对象。新对象的默认像素值为 RGBA(0,0,0,0)。在操作完成数组中的 color/alpha 信息之后，可以使用 putImageData()方法将图像数据拷贝回画布上。

createImageData()函数语法格式如表 2-51～表 2-52 所示。

以指定的尺寸（以像素计）创建新的 ImageData 对象：

表 2-51　getImageData()语法 1

JavaScript 语法	var imgData=*context*.createImageData(*width,height*);

创建与指定的另一个 ImageData 对象尺寸相同的新 ImageData 对象（不会复制图像数据）：

表 2-52　getImageData()语法 2

JavaScript 语法	var imgData=*context*.createImageData(*imageData*);

表 2-53 说明 createImageData()函数各参数的使用方法。

表 2-53　createImageData()参数

参数	描述
width	ImageData 对象的宽度，以像素计
height	ImageData 对象的高度，以像素计
imageData	另一个 ImageData 对象

ctx.drawImage(image,10,10); 表示为看出 createImageData()函数绘图效果不同，用 drawImage()函数从坐标(10,10)的位置绘制 image.jpg 图片；

var imgData=ctx.getImageData(20,20,100,100); 表示使用 getImageData()函数从画板中获取像素数据。

下面是针对 imgData01 像素数据的解析：

var imgData01=ctx.createImageData(imgData); 表示使用 createImageData()函数返回与 imgData 相同大小的 ImageData 对象；

```
        for (i=0; i<imgData01.width*imgData01.height*4;i+=4){
            imgData01.data[i+0]=255;
            imgData01.data[i+1]=0;
            imgData01.data[i+2]=0;
            imgData01.data[i+3]=255;
        }
```

使用循环对 imgData01 进行赋值；

ctx.putImageData(imgData01,10,110);

表示将所创建的像素数据 imgData01 画到 Canvas 画板以(10,110)为起始坐标的位置上。

下面是针对 imgData02 像素数据的解析：

var imgData02=ctx.createImageData(100,100);

表示使用 createImageData()函数返回一个大小为 100×100 的 ImageData 对象；

```
        for (i=0; i<imgData02.width*imgData02.height*4;i+=4){
            imgData02.data[i+0]=255;
```

```
        imgData02.data[i+1]=0;
        imgData02.data[i+2]=0;
        imgData02.data[i+3]=155;
    }
```

使用循环对 imgData02 进行赋值；

```
ctx.putImageData(imgData02,120,110);
```

表示将所创建的像素数据 imgData02 画到 Canvas 画板以(120,110)为起始坐标的位置上；运行效果如图 2-40 所示。

图 2-40　代码清单 2-36 运行后的效果图

第3章

CreateJS函数库

■ 第 2 章已经介绍 HTML5 Canvas 基础，虽然各大浏览器都支持 HTML5 Canvas，但是在使用 HTML5 开发游戏的时候，依然面临着几个大问题。第一，各浏览器对 JavaScript 和 HTML 的解析不一致，例如在 IE 中鼠标点击时的 offsetX 属性在 Firefox 中则必须用 layerX 来代替，这就使得在开发游戏的时候必须考虑代码在各个浏览器器重的兼容性问题，否则游戏将无法在所有的浏览器上都正常运行。第二，手机浏览器和 PC 浏览器也是有区别的，例如在手机浏览器支持 touch 事件，但是不支持 mouse 事件，而 PC 浏览器正好相反。第三，JavaScript 并非面向对象的编程，影响了代码的易读性。以上这些只是利用 Canvas 进行游戏开发时遇到的一部分问题，为了解决这个问题，最便捷的方式就是利用框架编程，因为一般来说，框架中会提供一些解决这类问题的方法。而 CreateJS 属于 JavaScript 的开源框架库，本章主要目的是认识基于 Canvas 的 CreateJS，学会使用 CreateJS 的函数库。

极客学院在线视频学习网址：
http://www.jikexueyuan.com/course/229_1.html
手机扫描二维码

认识 CreateJS

3.1 初识 CreateJS

3.1.1 下载 CreateJS

CreateJS 是一套可以构建丰富交互体验的 HTML5 游戏的开源工具包，旨在降低 HTML5 项目的开发难度和成本，让开发者以熟悉的方式打造更具现代感的网络交互体验。CreateJS 是一套 JS 库，是能配合 HTML5 一同（为网页）构建丰富交互体验的工具。本文的源码中附带有 CreateJS 库件的 0.7 版，用户也可以从 CreateJS 的官网上下载其最新版本，其下载地址为：http://www.createjs.com，如图 3-1 所示。

图 3-1 CreateJS 网站首页

CreateJS 套件分为四个库。

easeljs，这个是核心，包括了显示列表、事件机制；

preloadjs，用于预加载图片等；

tweenjs，用于控制元件的缓动；

soundjs，用于播放声音。

easeljs 81k，preloadjs 31k，soundjs 34k，tweenjs 18k，对于手机小动画或小游戏，其实只需要加载 easeljs 即可，因为核心库已经涵盖了简单图片的预加载功能。

3.1.2 介绍 CreateJS

关于 CreateJS，最关键是要理解图 3-2 所示的类关系图。

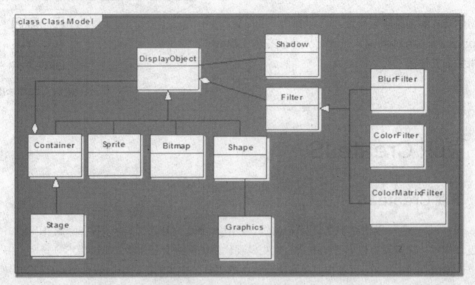

图 3-2　CreateJS 类关系图

图 3-3 列出了 CreateJS 的主要类结构，所有舞台上的内容都是元件，元件的基类是 DisplayObject。Container 可以包含子元件，舞台 Stage 本身也是 Container。另外 Sprite 用于表现 SpriteSheet 帧动画人物，Bitmap 用于展示纯静态的人物。

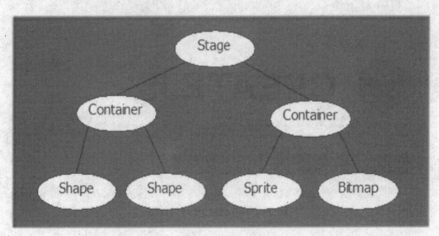

图 3-3　CreateJS 类结构

而 Filter 和 Shadow 则是滤镜分支，可以针对任意元件实现颜色变换、模糊、阴影等效果。使用滤镜的方式跟 Flash 一致，需要新建 Filter 实例，添加到目标元件的 FilterList 中，CreateJS 框架在下一帧就会把该元件加上滤镜效果。

开发步骤：
1. 下载类库，可以使用源代码，也可以使用压缩过的 min.js，就好像平时写网页一样。
2. 建立 html 和 canvas 标签，onload 后再执行 CreateJS 相关逻辑。

3. 编写 CreateJS 逻辑。这个非常简单，因为 CreateJS 只提供了最简单的基础功能，理解了基础功能后就可以叠积木了。

3.1.3 对比 CreateJS 与 Canvas

为了在案例中更好地理解和使用 CreateJS，本节利用案例比对的方式来演示 Canvas 与 CreateJS 实现蝴蝶飞翔的效果来说明用 CreateJS 开发动画效果的简便。

【案例 3-1】使用 Canvas 实现蝴蝶移动效果

代码清单 3-1

```html
<!DOCTYPE html>
<html>
<head>
<title></title>
</head>
<body style="margin: 20px">
<canvas id="canvas" width="700" height="400" style="border: black solid 1px"></canvas>
</body>
<script>
var c = document.getElementById("canvas");
var ctx = c.getContext("2d");
var butterfly = new Image();
    butterfly.src = "images/butterfly.png";
    butterfly.onload = drawButterflies;

function drawButterflies() {
ctx.drawImage(butterfly, 0, 0, 200, 138, 0, 0, 200, 138);
ctx.drawImage(butterfly, 0, 0, 200, 138, 200, 0, 200, 138);
ctx.drawImage(butterfly, 0, 0, 200, 138, 400, 0, 200, 138);
setTimeout(moveButterfly,1000);
    }
function moveButterfly(){
ctx.clearRect(0,0, c.width,c.height);
ctx.drawImage(butterfly, 0, 0, 200, 138, 0, 0, 200, 138);
ctx.drawImage(butterfly, 0, 0, 200, 138, 200, 200, 200, 138);
ctx.drawImage(butterfly, 0, 0, 200, 138, 400, 0, 200, 138);
    }
</script>
</html>
```

✧ 代码解析

var butterfly = new Image();
butterfly.src = "images/butterfly.png";
表示读取一个名为 images/butterfly.png 的图片；
butterfly.onload = drawButterflies;
表示给 Image 对象加了一个 onload 事件 drawButterflies；
drawButterflies 事件里面利用 drawImage 在 Canvas 画布上画出三个蝴蝶图像；
setTimeout(moveButterfly,1000);使用 setTimeout 函数间隔 1000 毫秒执行 moveButterfly 函数；
moveButterfly()函数先用 clearRect 擦除 Canvas 画板，然后使用 drawImage 函数重新绘制三个蝴

蝶图像的方式来实现蝴蝶移动的效果。

运行效果如图3-4所示。

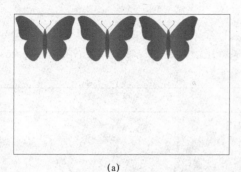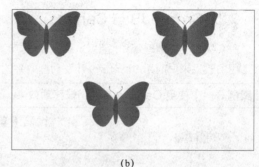

(a)　　　　　　　　　　　　　(b)

图3-4　使用Canvas实现蝴蝶移动效果

图3-4（a）是蝴蝶未移动之前的效果，图3-4（b）是蝴蝶移动后的效果。如果涉及的移动蝴蝶数量越来越多或者移动的效果越来越复杂，那么程序也越来越难以维护，【案例3-2】使用CreateJS中的EaselJS库能够清晰快捷地达到同样的效果，并且该代码的扩展性非常好。

【案例3-2】使用EaselJS实现蝴蝶移动效果

代码清单3-2

```
<!DOCTYPE html>
<html>
<head>
<title></title>
<script src="lib/easeljs-0.7.1.min.js"></script>
</head>
<body onload="init()">
<canvas id="canvas" width="1000" height="800" style="border: black solid 1px"></canvas>
</body>
<script>
var stage;
function init() {
stage = new createjs.Stage(document.getElementById('canvas'));
createjs.Ticker.addEventListener("tick", handleTick);
createjs.Ticker.setFPS(60);
start();
    }
function handleTick(e) {
stage.update();
    }
function start() {
drawButterflies();
    }
function drawButterflies() {
var imgPath = 'images/butterfly.png';
        butterfly1 = new createjs.Bitmap(imgPath);
        butterfly2 = new createjs.Bitmap(imgPath);
        butterfly3 = new createjs.Bitmap(imgPath);
        butterfly2.x = 200;
```

```
            butterfly3.x = 400;
    stage.addChild(butterfly1, butterfly2, butterfly3);
    stage.update();
    setTimeout(moveButterfly, 1000);
        }
        function moveButterfly() {
            butterfly2.y += 200;
        }
</script>
</html>
```

❖ 代码解析

<script src="lib/easeljs-0.7.1.min.js"></script>加载 easeljs-0.7.1.min.js 函数库；

<canvas id="canvas" width="1000" height="800" style="border: black solid 1px"></canvas>绘制 Canvas 画图板，所有的 CreateJS 函数操作都在整个画板上；

<body onload="init()">在 body 元素加了一个 onload 事件，该事件执行 init()操作；

var stage;定义画板变量；

init() 函数中的 stage = new createjs.Stage(document.getElementById('canvas'));
通过 createjs.Stage 获取 id 为"canvas" Canvas 画板来生成 Stage 舞台；

init() 函数中的 createjs.Ticker.addEventListener("tick", handleTick)
通过 createjs.Ticker.addEventListener 生成 tick 事件，事件执行 handleTick 函数，handleTick 函数 stage.update();实现的是舞台更新；

init() 函数中的 createjs.Ticker.setFPS(60)通过 createjs.Ticker.setFPS 为事件设置执行频率，该案例为 60 帧/秒；

init()函数中的 start()执行 drawButterflies()函数；

drawButterflies()函数中

var imgPath = 'images/butterfly.png';

读取一个名为 images/butterfly.png 的图片，

接着利用 createjs.Bitmap 创建三个蝴蝶对象 butterfly1、butterfly2、butterfly2，并且 butterfly2.x 设置蝴蝶的位置；

stage.addChild(butterfly1, butterfly2, butterfly3)将三个蝴蝶对象加载到 stage 舞台上；

stage.update();更新舞台；

setTimeout(moveButterfly, 1000);使用 setTimeout 函数间隔 1000 毫秒执行 moveButterfly 函数；

moveButterfly()函数先用 butterfly2.y += 200;通过改变 y 轴的方式来实现蝴蝶移动的效果。

运行效果如图 3-5 所示。

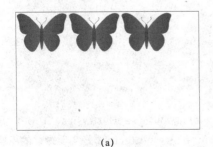

(a)　　　　　　　　　　　　　　(b)

图 3-5　使用 EaselJS 实现蝴蝶移动效果

图 3-5（a）是蝴蝶未移动之前的效果，图 3-5（b）是蝴蝶移动后的效果。图 3-5 与图 3-4 实现的效果一致，并且该代码的扩展性非常好，可以更方便地实现移动效果。

【案例 3-3】使用 EaselJS 实现更多蝴蝶移动效果

代码清单 3-3

```html
<!DOCTYPE html>
<html>
<head>
<title></title>
<script src="lib/easeljs-0.7.1.min.js"></script>
</head>
<body onload="init()" style="margin: 20px">
<canvas id="canvas" width="400" height="200" style="border: black solid 1px"></canvas>
</body>
<script>
    var stage;
    function init() {
        stage = new createjs.Stage(document.getElementById('canvas'));
        createjs.Ticker.addEventListener("tick", handleTick);
        createjs.Ticker.setFPS(60);
        start();
    }
    function handleTick(e) {
        stage.update();
    }
    function start() {
        drawButterflies();
    }
    function drawButterflies() {
        var imgPath = 'images/butterfly.png';
        butterfly1 = new createjs.Bitmap(imgPath);
        butterfly2 = new createjs.Bitmap(imgPath);
        butterfly2.x = 90;
        butterfly2.y = 40;
        stage.addChild(butterfly2, butterfly1);
        stage.update();
        setTimeout(swapButterfies, 1000);
    }
    function swapButterfies() {
        stage.swapChildren(butterfly1,butterfly2);
    }
</script>
</html>
```

❖ 代码解析

该案例与【案例 3-2】的代码基本相同，只是在 setTimeout(swapButterfies, 1000);调用的是 swapButterfies 函数，该函数调用 stage.swapChildren(butterfly1,butterfly2)来实现蝴蝶的更换效果，如图 3-6 所示。可以使用后面几节中的 CreateJS 函数来实现更多的舞台效果。

(a)　　　　　　　　　　　　　　(b)

图 3-6　使用 EaselJS 实现蝴蝶交换位置效果

3.2　CreateJS 包简介

CreateJS 库是 HTML5 游戏开发的引擎，掌握了 CreateJS 可以更方便地完成 HTML5 的游戏开发。本节分别使用一个案例来概要学习一下 CreateJS 提供的 EaselJS、TweenJS、SoundJS 和 PreLoadJS 函数库。

3.2.1　EaselJS 包

极客学院在线视频学习网址：
http://www.jikexueyuan.com/course/275_1.html
手机扫描二维码

CreateJS 介绍-EaselJS

【案例 3-4】使用 EaselJS 包实现舞台编写文字效果

代码清单 3-4

```
<!DOCTYPE html>
<html>
<head lang="en">
<meta charset="UTF-8">
<title></title>
<script src="easeljs-0.7.1.min.js"></script>
</head>
<body>
<canvas id="gamView" width="500px" height="500px" style="background-color:#cccccc"></canvas>
<script src="app.js"></script>
</body>
```

```
</html>
var stage = new createjs.Stage("gamView");
var text = new createjs.Text("Hello easeljs","36px Arial", "#777");
stage.addChild(text);
stage.update();
```

❖ 代码解析

index.html 定义页面框架，app.js 实现 CreateJS 编写文字效果，其中 var stage = new createjs.Stage("gamView")设置 CreateJS 舞台；

var text = new createjs.Text("Hello easeljs","36px Arial", "#777");设置 CreateJS 文字对象，文字格式为"36px Arial"，颜色为"#777"，文字内容为"Hello easeljs"；

stage.addChild(text);将文字对象添加到舞台；

stage.update();更新舞台将文字显示出来。

实现效果如图 3-7 所示。

图 3-7　CreateJS 文字效果

3.2.2　TweenJS 包

极客学院在线视频学习网址：
http://www.jikexueyuan.com/course/275_2.html
手机扫描二维码

CreateJS 介绍-TweenJS

【案例 3-5】使用 TweenJS 包实现图形变形效果

代码清单 3-5

```
<!DOCTYPE html>
<html>
<head lang="en">
```

```
<meta charset="UTF-8">
<title></title>
<script src="easeljs-0.7.1.min.js"></script>
<script src="tweenjs-0.6.1.min.js"></script>
</head>
<body>
<canvas id="gamView" width="500px" height="500px"
style="background-color:antiquewhite"></canvas>
<script src="app.js"></script>
</body>
</html>
var stage = new createjs.Stage("gamView");
stage.x = 100;
stage.y = 100;
var circle = new createjs.Shape();
circle.graphics.beginFill("#FF0000").drawCircle(0,0,50);
stage.addChild(circle);
createjs.Tween.get(circle,{loop:true})
.wait(1000)
.to({scaleX:0.2,scaleY:0.2})
.wait(1000)
.to({scaleX:1,scaleY:1},1000,createjs.Ease.bounceInOut)
createjs.Ticker.setFPS(20);
createjs.Ticker.addEventListener("tick",stage);
```

◆ 代码解析

Index.html 实现网页布局效果，其中<script src="easeljs-0.7.1.min.js"></script><script src="tweenjs-0.6.1.min.js"></script>加载 easeljs-0.7.1.min.js 和 tweenjs-0.6.1.min.js 函数库；

app.js 中 var stage = new createjs.Stage("gamView");创建舞台对象；

stage.x = 100;stage.y = 100;设置舞台的初始位置为（100,100）；

var circle = new createjs.Shape();设置舞台画板；

circle.graphics.beginFill("#FF0000").drawCircle(0,0,50);用一条长语句在画板上绘制颜色为"#FF0000"、以（0,0）为圆心半径为 50 的圆形；

stage.addChild(circle);将圆形加载到画板上；

接着使用长语句实现图像的动态效果，首先用 createjs.Tween.get(circle,{loop:true})设置 circle 的循环效果为 true；

.wait(1000)等待 1000 毫秒；

.to({scaleX:0.2,scaleY:0.2})将 X、Y 轴都缩放为 0.2；

.wait(1000)等待 1000 毫秒；

.to({scaleX:1,scaleY:1},1000,createjs.Ease.bounceInOut)使 X、Y 轴都恢复到原状态；createjs.Ease.bounceInOut 实现渐进渐出效果；

最后 createjs.Ticker.setFPS(20)设置刷新频率为 20 帧/秒；

createjs.Ticker.addEventListener("tick",stage);添加舞台 tick 事件，事件为更新舞台的操作。

代码实现效果如图 3-8 所示。

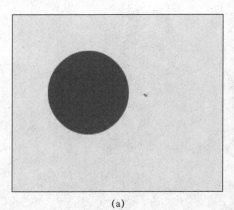
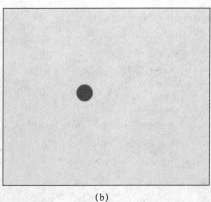

(a)   (b)

图 3-8   图形变形效果

3.2.3   SoundJS 包

极客学院在线视频学习网址：
http://www.jikexueyuan.com/course/275_3.html
手机扫描二维码

CreateJS 介绍-SoundJS

【案例 3-6】 使用 SoundJS 包载入音频效果

代码清单 3-6

```html
<!DOCTYPE html>
<html>
<head lang="en">
<meta charset="UTF-8">
<title></title>
<script src="soundjs-0.6.1.min.js"></script>
</head>
<body>
<div>
<h1 id="status">startup </h1>
</div>
<script src="app.js"></script>
</body>
</html>
var displayStatus;
displayStatus = document.getElementById("status");
src = "1.mp3";
createjs.Sound.alternateExtensions=["mp3"];
createjs.Sound.addEventListener("fileload",playSound);
createjs.Sound.registerSound(src);
```

```
displayStatus.innerHTML = "Waiting for load to complete";
function playSound(event){
    sounceIntance = createjs.Sound.play(event.src);
   displayStatus.innerHTML = "Playing Source:" + event.src;
}
```

✧ 代码解析

Index.html 实现网页布局效果，其中<script src="soundjs-0.6.1.min.js"></script>加载 soundjs-0.6.1.min.js 函数库；

app.js 绘制 CreateJS 动画效果；

displayStatus = document.getElementById("status");获取 id 为"status"的 h1 对象；

src = "1.mp3"设置 mp3 音乐的地址为 1.mp3；

createjs.Sound.alternateExtensions=["mp3"];设置播放的资源类型为 mp3 格式；

displayStatus.innerHTML = "Waiting for load to complete";设置 h1 的初始状态文字为："Waiting for load to complete"；

createjs.Sound.registerSound(src);对资源进行加载与注册；

createjs.Sound.addEventListener("fileload",playSound);设置一个 fileload 的资源准备事件，资源准备好后触发 playSound 函数，playSound(event)函数 createjs.Sound.play(event.src);来播放资源 1.mp3，displayStatus.innerHTML = "Playing Source:" + event.src;在 h1 中显示播放音乐的资源名称。

代码实现效果如图 3-9～图 3-11 所示。

startup

图 3-9 初始未加载图效果

Waiting for load to complete

图 3-10 正在加载效果

Playing Source:1.mp3

图 3-11 开始播放效果

3.2.4 PreLoadJS 包

极客学院
jikexueyuan.com

极客学院在线视频学习网址：
http://www.jikexueyuan.com/course/275_4.html
手机扫描二维码

CreateJS 介绍-PreLoadJS

【案例 3-7】使用 PreLoadJS 包实现图片载入

代码清单 3-7

```html
<!DOCTYPE html>
<html>
<head lang="en">
<meta charset="UTF-8">
<title></title>
<script src="preloadjs-0.6.1.min.js"></script>
<style>
    .image{
        max-width:320px;
         max-height: 240px;
        border:1px solid #555;
        margin:5px;
    }
</style>
</head>
<body>
<h1 id="status">startup </h1>
<img class="image" id="texas">
<img class="image" id="bluebird">
<img class="image" id="nepal">
<script src="app.js"></script>
</body>
</html>
var preload;
preload = new createjs.LoadQueue(false,"assets/");
var plugin={
    getPreloadHandlers:function(){
        return{
            types:["image"],
            callback:function(src){
                var id = src.toLowerCase().split("/").pop().split(".")[0];
                var img = document.getElementById(id);
                window.alert(img.getAttribute("id"));
                return {tag:img};
            }
        }
    }
}
preload.installPlugin(plugin);
preload.loadManifest([
    "Autumn.png",
    "BlueBird.png",
    "Nepal.jpg",
    "Texas.jpg"
]);
```

◆ 代码解析

Index.html 实现网页布局效果，其中<script src=" preloadjs-0.6.1.min.js "></script>加载 preloadjs-0.6.1.min.js 函数库；<img>标签加载三个图片；<style>标签设置的 css 显示效果；

app.js 中通过 createjs.LoadQueue(false,"images/");来加载图片资源，getPreloadHandlers 来获取图片名称，其中图片是 types 为["image"]class 类型，id 经过 src.toLowerCase().split("/").pop().split(".")[0];处理来获取。

接在通过 preload.installPlugin(plugin)函数来触发事件，通过 preload.loadManifest 来加载每个图片的资源。

代码显示效果如图 3-12 所示。

图 3-12　图片载入效果

3.3　EaselJS 基础

EaselJS 用来处理 HTML5 的 Canvas，掌握了 EaselJS 的容器、绘图和事件，就具有了制作 HTML5 小游戏的能力。

3.3.1　EaselJS 容器

一个舞台（Stage）是根层次 Container 的一个显示列表。每一次 tick 方法被调用，它会将其显示列表呈现给它的目标画布。

语法格式：创建一个舞台，并增加一个子元素到里面，然后使用 Ticker 更新的子元素和使用 update 重画舞台。

极客学院在线视频学习网址：
http://www.jikexueyuan.com/course/276_1.html
手机扫描二维码

CreateJS 基础-EaselJS 容器

代码清单 3-8

```
var stage = new createjs.Stage("canvasElementId");
var image = new createjs.Bitmap("imagePath.png");
stage.addChild(image);
createjs.Ticker.addEventListener("tick", handleTick);
function handleTick(event) {
    image.x += 10;
```

```
stage.update();
}
```
Stage 设计的源代码在下载的包文件..\\ EaselJS-0.8.1\src\easeljs\display\Stage.js 中。

代码清单 3-9

```
function Stage(canvas) {
         this.Container_constructor();
```

其相关的方法、属性、事件可以查看附件的 API 文档。

Stage 是展示列表容器的最高层（这个 stage 就是 canvas 元素）。在它的下层就是 HTML 片段和相关用于驱动 EaselJS 的 JS 代码。本节主要讲解 EaselJS 的容器，让学员掌握 EaselJS 中容器的操作。

【案例 3-8】用 CreateJS Stage()实现画图效果

代码清单 3-10

```
<!DOCTYPE html>
<html>
<head lang="en">
<meta charset="UTF-8">
<title></title>
<script src="easeljs-0.7.1.min.js"></script>
</head>
<body>
<canvas id="gamView" width="400px" height="400px"
style="background-color:#cccccc"></canvas>
<script src="app.js"></script>
</body>
</html>
var stage = new createjs.Stage("gamView");
stage.alpha = 0.5;
stage.x = 100;
stage.y = 100;
stage.scaleX = 2;
stage.scaleY = 0.5;
var Rect = new createjs.Shape();
Rect.graphics.beginFill("#ff0000");
Rect.graphics.drawRect(5,5,50,50);
stage.addChild(Rect);
stage.update();
```

❖ 代码解析

Index.html 实现网页布局效果，其中<script src="easeljs-0.7.1.min.js"></script>加载 easeljs-0.7.1.min.js 函数库；

app.js 中 var stage = new createjs.Stage("gamView");创建舞台对象；

stage.alpha = 0.5;设置舞台的透明度为 0.5;

stage.x = 100;stage.y = 100;设置舞台的初始位置为（100,100）;

stage.scaleX = 2;stage.scaleY = 0.5;设置图形 X 轴增大 2 倍，Y 轴缩略显示为 0.5 倍；

var circle = new createjs.Shape();设置舞台画板；

Rect.graphics.beginFill("#ff0000");设置画板颜色为"#FF0000";

Rect.graphics.drawRect(5,5,50,50);在坐标为（5,5）的位置画出长宽为 50 的矩形；

stage.addChild(Rect);将矩形加载到画板上；

stage.update();更新舞台。

代码显示效果如图 3-13 所示。

图 3-13　Stage 画图效果

容器是一个嵌套显示列表，可以利用组合的方式来显示元素。例如，可以将手臂、腿部、躯干和头部实例聚在一起，把它们转换为一组，同时还可以将各个部分相对彼此移动。容器中各个元素称之为容器的孩子，孩子类的.transform 和 alpha 属性与它们的父容器连接。

例如，容器（X = 50 和 Alpha = 0.7）中有一个 Shape（X = 100 和 Alpha = 0.5），最终画布的效果将是 X = 150 和 Alpha = 0.35。定义一个容器需要开销，所以通常不应该创建一个容器来容纳一个孩子。

Stage 相关的设计的源代码在下载的包文件

..\\ EaselJS-0.8.1\src\easeljs\display\Container.js 中。

代码清单 3-11

```
function Container() {
        this.DisplayObject_constructor();
```

其相关的方法、属性、事件可以查看附件的 API 文档。

【案例 3-9】CreateJS Container 实现图形移动效果

代码清单 3-12

```
<!DOCTYPE html>
<html>
<head lang="en">
<meta charset="UTF-8">
<title></title>
<script src="easeljs-0.7.1.min.js"></script>
<script src="ChildContainer.js"></script>
</head>
<body>
<canvas id="gamView" width="400px" height="400px"
style="background-color:#cccccc"></canvas>
<script src="container.js"></script>
</body>
</html>
```

container.js

代码清单 3-13

```
var stage = new createjs.Stage("gamView");
var gameView = new createjs.Container();
stage.addChild(gameView);
var c = new ChildContainer();
gameView.addChild(c);
stage.update();
```

ChildContainer.js

代码清单 3-14

```
function ChildContainer(){
    var Rect = new createjs.Shape();
    Rect.graphics.beginFill("#ff0000");
    Rect.graphics.drawRect(5,5,50,50);
    Rect.graphics.endFill();;
    var Text = new createjs.Text("1","30px Arial", "#ffffff");
    this.addChild(Rect);
    this.addChild(Text);
}
ChildContainer.prototype = new createjs.Container();
```

Index.html 实现网页布局效果，其中<script src="easeljs-0.7.1.min.js"></script>加载 easeljs-0.7.1.min.js 函数库；

<script src="ChildContainer.js"></script><script src="container.js"></script>加载 ChildContainer.js 和 container.js 源代码；

container.js 中 var gameView = new createjs.Container();stage.addChild(gameView);定义一个容器，并把容器添加到舞台上面；

var c = new ChildContainer();调用 ChildContainer.js 包中的 ChildContainer()函数；

Rect.graphics.beginFill("#ff0000"); Rect.graphics.drawRect(5,5,50,50);在（5,5）位置画出颜色为"#ff0000"、长宽为（50×50）的矩形；

var Text = new createjs.Text("1","30px Arial", "#ffffff");定义字形为"30px Arial"、颜色为"#ffffff"、内容为"1"的字体；

this.addChild(Rect); this.addChild(Text);现将矩形放入容器，再将字体放入容器，形成层叠效果；

ChildContainer.prototype = new createjs.Container();创建容器的原型；

接着回到主函数 container.js 中 gameView.addChild(c);stage.update();把容器元素添加到容器中，并对舞台进行更新显示。

代码显示效果如图 3-14 所示。

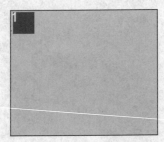

图 3-14 CreateJS 文字效果

如图 3-15 所示。将代码块 var Text = new createjs.Text("1","30px Arial", "#ffffff")的文字内容修改为 Hello：var Text = new createjs.Text("Hello","30px Arial", "#ffffff")；就可以显示其他文字；

添加代码块：gameView.x = 100；gameView.y = 100；就可以轻松实现实现文字与容器的一起移动的效果。

图 3-15　文字容器一起移动效果

【案例 3-10】CreateJS Container 实现更多移动效果

代码清单 3-15

```
<!DOCTYPE html>
<html>
<head>
<title></title>
<script src="easeljs-0.7.1.min.js"></script>
<script src="tweenjs-0.5.1.min.js"></script>
</head>
<body onload="init()">
<canvas id="canvas" width="600" height="400" style="border: black solid 1px"></canvas>
</body>
<script>
    var stage;
    function init() {
        stage = new createjs.Stage(document.getElementById('canvas'));
        createjs.Ticker.setFPS(60);
        createjs.Ticker.addEventListener("tick", function (e) {
            stage.update();
        });
        // sample();
        // sample2();
        sample3();
    }
    function sample() {
        var container1 = new createjs.Container();
        var container2 = new createjs.Container();
        var pepper = new createjs.Bitmap('img/pepper.png')
        var circle = new createjs.Shape(new createjs.Graphics().beginFill('#FF0000').drawCircle(0, 0, 50));
        var square = new createjs.Shape(new createjs.Graphics().beginFill('#00FF00').drawRect(0, 0, 50, 50));
        var txt = new createjs.Text("Hello Containers", "20px Arial", "#000");
```

```
            var bg = new createjs.Shape(new createjs.Graphics().beginStroke('#000').drawRect(0, 0, 250, 250));
            container1.addChild(bg);
            bg = new createjs.Shape(new createjs.Graphics().beginStroke('#000').drawRect(0, 0, 250, 250));
            container2.addChild(bg);
            txt.x = txt.y = 10;
            circle.x = circle.y = 125;
            container1.addChild(txt, circle);
            square.x = square.y = 10;
            pepper.x = pepper.y = 100;
            container2.addChild(square, pepper);
            container1.x = 20;
            container2.x = 320;
            container1.y = container2.y = 40;
            stage.addChild(container1, container2);
            stage.update();
        }
        function sample2() {
            var container = new createjs.Container();
            var pepper = new createjs.Bitmap('img/pepper.png')
            var txt = new createjs.Text("Green Pepper", "20px Arial", "#000");
            var bg = new createjs.Shape(new createjs.Graphics().beginStroke('#000').drawRect(0, 0, 250, 250));
            txt.x = txt.y = 10;
            pepper.x = pepper.y = 80;
            container.regX = container.regY = 125;
            container.x = 300;
            container.y = 200;
            container.addChild(bg, txt, pepper);
            stage.addChild(container);
            createjs.Tween.get(container).to({rotation:360},4000);
        }
        function sample3(){
            var container = new createjs.Container();
            var pepper = new createjs.Bitmap('img/pepper.png')
            var txt = new createjs.Text("Green Pepper", "20px Arial", "#000");
            var bg = new createjs.Shape(new createjs.Graphics().beginStroke('#000').drawRect(0, 0, 250, 250));
            txt.x = txt.y = 10;
            pepper.x = pepper.y = 80;
            container.regX = container.regY = 125;
            container.x = 150;
            container.y = 200;
            container.addChild(bg, txt, pepper);
            container2 = container.clone(true);
            container2.x = 430;
            container2.y = 200;
            stage.addChild(container,container2);
        }
        function startGame() {

        }
    </script>
</html>
```

◆ 代码解析

本案例使用三个实例 sample1、sample2、sample3 来实现容器的更多移动效果。

sample()中 var container1 = new createjs.Container(); var container2 = new createjs.Container();定义两个容器；

var pepper = new createjs.Bitmap('img/pepper.png')载入图片；

circle、square、txt 分别定义一个圆形、方形和文字元素；

定义一个 bg 元素并使用 container1.addChild(bg);将 bg 添加到 container1 上；

bg = new createjs.Shape(new createjs.Graphics().beginStroke('#000').drawRect(0, 0, 250, 250));再修改 bg 元素并使用 container2.addChild(bg);将 bg 添加到 container2 上；

txt.x = txt.y = 10; circle.x = circle.y = 125; container1.addChild(txt, circle);调整 txt 与 Circle 位置并将其分别加到 container1 上；

square.x = square.y = 10; pepper.x = pepper.y = 100; container2.addChild(square, pepper);调整 square 与 pepper 位置并将其分别加到 container2 上；

container1.x = 20；container2.x = 320；container1.y = container2.y = 40;添加 container1、container2 的位置；

stage.addChild(container1，container2); stage.update();将 container1、container2 添加到画板上并刷新画板。

代码显示效果如图 3-16 所示。

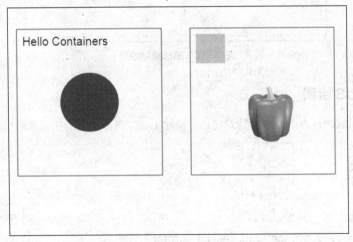

图 3-16　更多图形移动效果 1

sample2()中 container.addChild(bg，txt，pepper);将定义的三个元素 bg、txt、pepper 分别加入画板中，然后利用 Tween 包中的 createjs.Tween.get(container).to({rotation:360},4000);实现旋转效果。

代码显示效果如图 3-17 所示。

sample3()中 container.addChild(bg，txt，pepper);将定义的三个元素 bg、txt、pepper 分别加入画板中，然后利用 container2 = container.clone(true);克隆一个相同的容器，再用 container2.x = 430; container2.y = 200;修改容器的位置。

代码显示效果如图 3-18 所示。

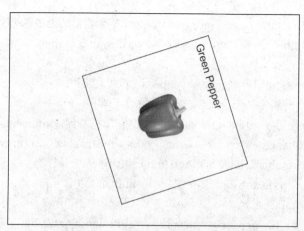

图 3-17　更多图形移动效果 2

图 3-18　图形复制效果

3.3.2　EaselJS 绘图

本节主要讲解 EaselJS 的绘图，让学员可以在画布上绘制自己想要的任何图形和效果。

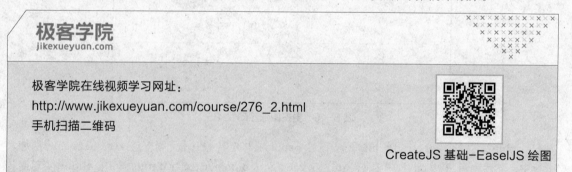

极客学院在线视频学习网址：
http://www.jikexueyuan.com/course/276_2.html
手机扫描二维码

CreateJS 基础-EaselJS 绘图

1. Graphics Class

图形类公开了一个易于使用的接口，用于生成矢量绘图指令，并将其绘制到指定的上下文中。用户可以使用 easeljs 框架来调用 draw 类型的绘制，也可以使用 Shape 对象绘制矢量图形。

代码清单 3-16 所示，有两种方法来处理图形对象：一是图形实例调用方法（Graphics API），二是以图形命令对象并通过 append 方法将它们添加到图形队列。前者简化了开始和结束路径、填充和笔画编程。

代码清单 3-16

```
var g = new createjs.Graphics();
g.setStrokeStyle(1);
g.beginStroke("#000000");
g.beginFill("red");
g.drawCircle(0,0,30);
```

图形实例中的所有绘图方法被链接在一起。例如，代码清单 3-17 将生成一个以红色的笔画和蓝色填充来绘制矩形的指令。

代码清单 3-17

```
myGraphics.beginStroke("red").beginFill("blue").drawRect(20, 20, 100, 50);
```

每个图形接口调用生成一个命令对象。创建可访问的最后一个命令可以通过 command 命令：

代码清单 3-18

```
var fillCommand = myGraphics.beginFill("red").command;
    // 稍后... 更新填充样式/颜色：
    fillCommand.style = "blue";
    // 或者将其更改为一个位图填充：
fillCommand.bitmap(myImage);
```

更直接地控制渲染，可以实例化和直接将附加命令对象添加到图形队列中去。在这种情况下，需要手动管理路径创建，并确保将 fill/stroke 应用到定义路径中去。

代码清单 3-19

```
// 开始一个新的路径，graphics.begincmd是一个可重用的BeginPath实例：
myGraphics.append(createjs.Graphics.beginCmd);
    // 我们需要在应用fill前定义路径：
var circle = new createjs.Graphics.Circle(0,0,30);
myGraphics.append(circle);
    // fill我们刚刚定义的路径：
var fill = new createjs.Graphics.Fill("red");
myGraphics.append(fill);
```

这些方法可以一起使用，如插入一个自定义命令。

代码清单 3-20

```
myGraphics.beginFill("red");
var customCommand = new CustomSpiralCommand(etc);
myGraphics.append(customCommand);
myGraphics.beginFill("blue");
myGraphics.drawCircle(0, 0, 30);
```

表 3-1 所示是图形类缩略函数，这是用一个或两个字母的方法来表示所有的图形方法的快捷方式。这些紧凑的指令通过 CreateJS 工具包来生成代码的可读性。所有的小方法被标记为受保护的，具体的使用方法可以查阅文档。

表 3-1 图形类缩略函数

Tiny	Method	Tiny	Method
mt	moveTo	lt	lineTo
a/at	arc/arcTo	bt	bezierCurveTo
qt	quadraticCurveTo(also curveTo)	r	rect
cp	closePath	c	clear

续表

Tiny	Method	Tiny	Method
f	beginFill	lf	beginLinearGradientFill
rf	beginRadialGradientFill	bf	beginBitmapFill
ef	endFill	ss/ sd	setStrokeStyle/setStrokeDash
s	beginStroke	ls	beginLinearGradientStroke
rs	beginRadialGradientStroke	bs	beginBitmapStroke
es	endStroke	dr	drawRect
rr	drawRoundRect	rc	drawRoundRectComplex
dc	drawCircle	de	drawEllipse
dp	drawPolyStar	p	decodePath

代码清单 3-21 使用表 3-1 的缩略命令实现了绘制红色线条、蓝色背景的矩形。

代码清单 3-21

```
myGraphics.s("red").f("blue").r(20, 20, 100, 50);
```

【案例 3-11】使用 Graphics 实现画图效果

代码清单 3-22

```
<!DOCTYPE html>
<html>
<head lang="en">
<meta charset="UTF-8">
<title></title>
<script src="easeljs-0.7.1.min.js"></script>
</head>
<body>
<canvas id="gamView" width="400px" height="400px"
style="background-color:#cccccc"></canvas>
<script src="app.js"></script>
</body>
</html>
var stage = new createjs.Stage("gamView");
var gameView = new createjs.Container();
stage.addChild(gameView);

var Rect = new createjs.Shape();
Rect.graphics.beginFill("#ff00ff");
Rect.graphics.drawRect(0,0,100,50);
gameView.addChild(Rect);

var Ellipse = new createjs.Shape();
Ellipse.graphics.beginFill("#00ff00");
Ellipse.graphics.drawEllipse(100,100,100,80);
gameView.addChild(Ellipse);

stage.update();
```

Index.html 实现网页布局效果，<script src=" app.js"></script>加载 app.js 源代码；

Rect 在(0,0)位置绘制一个长宽为(100,50)紫色的矩形；
Ellipse 在(100,100)位置绘制一个最长直径为 100、最高直径为 80 的绿色椭圆；
代码显示效果如图 3-19 所示。

图 3-19　Graphics 画图效果

修改 Index.html，添加<script src="Circle.js"></script>加载 Circle.js 源代码；
并修改 app.js 为：

代码清单 3-23

```
var stage = new createjs.Stage("gamView");
var gameView = new createjs.Container();
stage.addChild(gameView);

gameView.x = 100;
gameView.y = 100;

var c= new Circle();
c.setCircleType(1);
stage.addChild(c);
stage.update();
```

var c= new Circle();c.setCircleType(1);调用 Circle.js 中的函数实现绘制圆形与设置不用颜色的效果。Circle.js 如下：

代码清单 3-24

```
function Circle(){
    createjs.Shape.call(this);
    this.setCircleType = function(type){
        this._circleType = type;
        switch (type){
            case Circle.TYPE_RED:
                this.setColor("#ff0000");
                break;
            case Circle.TYPE_GREEN:
                this.setColor("#00ff00");
                break;
```

```
                }
            }
            this.setColor = function(color){
                this.graphics.beginFill(color);
                this.graphics.drawCircle(100,100,50);
                this.graphics.endFill();
            }
        }
        Circle.prototype = new createjs.Shape();
        Circle.TYPE_RED = 1;
        Circle.TYPE_GREEN = 2;
```

Circle()函数利用一个 switch 判断实现设置颜色的效果,当 setCircleType(1)调用的参数值为 1 时,画出的圆形为红色,当 setCircleType(2)调用的参数值为 2 时,画出的圆形为绿色,这样可以灵活地去设计颜色内容。

代码显示效果如图 3-20 所示。

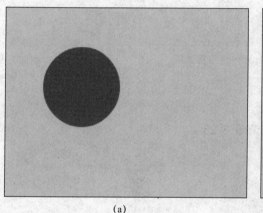
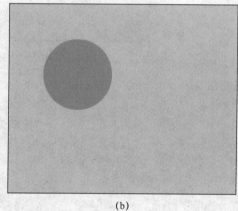

(a)　　　　　　　　　　　　　(b)

图 3-20　用属性值设置来画图

2. Shape

Shape 元素可以在显示列表中显示矢量艺术。它包含 Graphics 的实例,公开了所有的矢量绘图方法。Graphics 实例可以在多个 Shape 实例之间共享,以显示相同却具有不同位置或变换的矢量图形。

用法:

代码清单 3-25

```
var graphics = new createjs.Graphics().beginFill("#ff0000").drawRect(0, 0, 100, 100);
var shape = new createjs.Shape(graphics);

//也可以如上面一样交替使用Shape类的图形属性进行渲染。
var shape = new createjs.Shape();
shape.graphics.beginFill("#ff0000").drawRect(0, 0, 100, 100);
```

【案例 3-12】使用 Shape 实现各种图形效果

代码清单 3-26

```
<!DOCTYPE html>
<html>
<head>
<title></title>
```

```html
<script src="lib/easeljs-0.7.1.min.js"></script>
<script src="lib/tweenjs-0.5.1.min.js"></script>
</head>
<body onload="init()" style="margin: 20px">
<canvas id="canvas" width="800" height="600" style="border: black solid 1px"></canvas>
</body>
<script>
    var stage;
    function init() {
        stage = new createjs.Stage(document.getElementById('canvas'));
        createjs.Ticker.addEventListener("tick", handleTick);
        createjs.Ticker.setFPS(60);
        start();
    }
    function handleTick(e) {
        stage.update();
    }

    function start() {
        //RECTANGLE
        var rectangle = new createjs.Shape();
        rectangle.graphics.beginStroke('#000');
        rectangle.graphics.beginFill('#FF0000');
        rectangle.graphics.drawRect(0, 0, 150, 100);
        rectangle.x = rectangle.y = 20;
        stage.addChild(rectangle);
        //CIRCLE
        var circle = new createjs.Shape();
        circle.graphics.beginStroke('#000');
        circle.graphics.beginFill('#FFF000');
        circle.graphics.drawCircle(0, 0, 50);
        circle.x = 250;
        circle.y = 70;
        stage.addChild(circle);
        //STAR
        var poly = new createjs.Shape();
        poly.graphics.beginStroke('#000');
        poly.graphics.beginFill('#90ABC2');
        poly.graphics.drawPolyStar(0, 0, 60, 6, 0.6);
        poly.x = 400;
        poly.y = 70;
        stage.addChild(poly);
        //TRIANGLE
        var tri = new createjs.Shape();
        tri.graphics.beginStroke('#000');
        tri.graphics.beginFill('#00FF00');
        tri.graphics.moveTo(50, 0)
            .lineTo(0, 100)
            .lineTo(100, 100)
            .lineTo(50, 0);
        tri.x = 20;
```

```
            tri.y = 150;
            stage.addChild(tri);
            ///ROUNDED RECTANGLE
            var roundedRectangle = new createjs.Shape();
            roundedRectangle.graphics.beginStroke('#000');
            roundedRectangle.graphics.beginFill('#F7D0D1');
            roundedRectangle.graphics.drawRoundRect(0,0,400,100,10);
            roundedRectangle.x = roundedRectangle.y = 150;
            stage.addChild(roundedRectangle);
        }
    </script>
</html>
```

❖ 代码解析

function handleTick(e) { stage.update(); }实现舞台更新效果；

createjs.Ticker.addEventListener("tick", handleTick); createjs.Ticker.setFPS(60);设置更新的频率为 60 帧/秒；

调用 start();函数绘制 RECTANGLE 矩形、CIRCLE 圆形、STAR 星形、TRIANGLE 三角形、ROUNDED RECTANGLE 圆角矩形。

代码显示效果如图 3-21 所示。

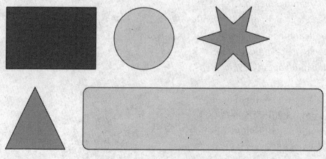

图 3-21 各种不同图形效果

3. 图形变换

图形的旋转：

【案例 3-13】使用 CreateJS 实现旋转效果

代码清单 3-27

```
<!DOCTYPE html>
<html>
<head>
<title></title>
<script src="lib/easeljs-0.7.1.min.js"></script>
<script src="lib/tweenjs-0.5.1.min.js"></script>
</head>
<body onload="init()" style="margin: 20px">
<canvas id="canvas" width="400" height="300" style="border: black solid 1px"></canvas>
</body>

<script>
```

```
        var stage;
        function init() {
            stage = new createjs.Stage(document.getElementById('canvas'));
            createjs.Ticker.addEventListener("tick", handleTick);
            createjs.Ticker.setFPS(60);
            start();
        }
        function handleTick(e) {
            stage.update();
        }
        function start() {
            var g = new createjs.Graphics();
            g.beginStroke('#000').beginFill('#FF0000').drawRect(0, 0, 100, 100);
            var square = new createjs.Shape(g);
            square.x = 150;
            square.y = 100;
            stage.addChild(square);
            createjs.Tween.get(square).to({rotation:360},3000);
        }
    </script>
</html>
```

❖ 代码解析

createjs.Tween.get(square).to({rotation:360},3000);实现矩形沿着顶角位置（150,100）旋转360度。

代码显示效果如图3-22所示（动态效果请运行源代码）。

图3-22 旋转效果

修改代码段：

代码清单 3-28

```
            var g = new createjs.Graphics();
            g.beginStroke('#000').beginFill('#FF0000').drawRect(0, 0, 100, 100);
            var square = new createjs.Shape(g);
square.regX = square.regY = 50;
square.x = stage.canvas.width / 2;
square.y = stage.canvas.height / 2;
stage.addChild(square);
            createjs.Tween.get(square).to({rotation:360},3000);
```

square.regX = square.regY = 50; square.x = stage.canvas.width / 2; square.y = stage.canvas.height / 2; 改变旋转方式为绕着中心点旋转。

代码显示效果如图 3-23 所示（动态效果请运行源代码）。

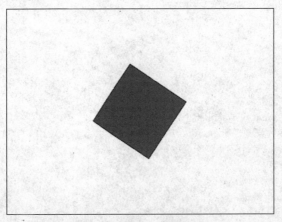

图 3-23 绕中心旋转效果

【案例 3-14】使用 CreateJS 实现小球碰撞效果

代码清单 3-29

```html
<!DOCTYPE html>
<html>
<head>
<title></title>
<script src="../lib/easeljs-0.7.1.min.js"></script>
</head>
<body style="margin: 20px">
<canvas id="canvas" width="800" height="600" style="border: black solid 1px"></canvas>
</body>
<script>
    var stage = new createjs.Stage(document.getElementById('canvas'));
    var direction = 1;
    var velocity = 10;

    var circle = new createjs.Shape();
    circle.graphics.beginStroke('#000');
    circle.graphics.beginFill('#FFF000');
    circle.graphics.drawCircle(0, 0, 50);
    circle.radius = 50;
    circle.x = 100;
    circle.y = 300;

    stage.addChild(circle);

    createjs.Ticker.addEventListener("tick", tick);
    createjs.Ticker.setFPS(60);

    function updateCircle() {
```

```
            var nextX = circle.x + (velocity * direction);
            if (nextX > stage.canvas.width − circle.radius) {
                nextX = stage.canvas.width − circle.radius;
                direction *= −1;
            }
            else if (nextX < circle.radius) {
                nextX = circle.radius;
                direction *= −1;
            }
            circle.nextX = nextX;
        }
        function renderCircle() {
            circle.x = circle.nextX;
        }

        function tick(e) {
            updateCircle();
            renderCircle();
            stage.update();
        }

</script>
</html>
```

❖ 代码解析

Circle 先在舞台上（100，300）的位置画上半径为 50 的黄色圆形；

createjs.Ticker.addEventListener("tick", tick); createjs.Ticker.setFPS(60);以 60 帧/秒的速度调用 tick 函数；

tick(e)函数包含三个动作，分别为 updateCircle()、renderCircle()、stage.update()。

其中 updateCircle()判断圆形运动的方向，默认的运动方向为向右运动，如果 nextX > stage.canvas.width − circle.radius，则撞到右边的边界，运动方向改为向左运动；如果 nextX < circle.radius，则撞到左边的边界，运动方向改为向右运动；

circle.nextX = nextX;确定圆形运动的下一次位置；

renderCircle() 函数中的 circle.x = circle.nextX;将圆形运动的下一次位置交给 circle 对象的 x 属性。

经过 60 帧/秒的调用 tick 函数来实现圆形不断碰撞弹跳的效果。

代码显示效果如图 3-24 所示（动态效果请运行源代码）。

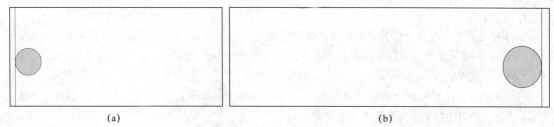

图 3-24　小球边缘碰撞效果

3.3.3 EaselJS 事件

本节主要讲解 EaselJS 事件，让学员可以在画布上进行操作，实现对事件的监听和处理。

极客学院在线视频学习网址：
http://www.jikexueyuan.com/course/276_3.html
手机扫描二维码

CreateJS 基础-EaselJS 事件

【案例 3-15】使用 CreateJS 实现鼠标事件 1

代码清单 3-30

```
var stage = new createjs.Stage("gamView");
var gameView = new createjs.Container();
stage.addChild(gameView);

var Rect = new createjs.Shape();
Rect.graphics.beginFill("#ff00ff");
Rect.graphics.drawRect(0,0,100,100);
gameView.addChild(Rect);
stage.update();

//Rect.addEventListener("click",function(){
//    alert("点了图形");
//});

//Rect.addEventListener("click",function(e){
//    alert("X=" + e.localX + "--Y=" + e.localY);
//});

//Rect.addEventListener("dblclick",function(e){
//    alert("X=" + e.localX + "--Y=" + e.localY);
//});
```

✧ 代码解析

Rect 在位置(0,0)绘制一个长宽为(100,100)的紫色矩形；Rect.addEventListener("click",function(){ alert("点了图形");});添加鼠标单击事件，事件为弹出内容为"点了图形"的对话框。

代码显示效果如图 3-25 所示。

Rect.addEventListener("click",function(e){ alert("X=" + e.localX + "--Y=" + e.localY);});添加鼠标单击事件，事件为弹出内容为 X 轴、Y 轴左边的对话框；代码显示效果如图 3-26 所示。

图 3-25 鼠标单击事件 1

图 3-26 鼠标单击显示坐标

Rect.addEventListener("dblclick",function(e){ alert("X=" + e.localX + "--Y=" + e.localY);}); 添加鼠标双击事件，事件为弹出内容为 X 轴、Y 轴左边的对话框。

代码显示效果如图 3-27 所示。

图 3-27 鼠标双击显示坐标

【案例 3-16】使用 CreateJS 实现鼠标事件 2

代码清单 3-31

```
<!DOCTYPE html>
<html>
<head>
<title></title>
<script src="../lib/easeljs-0.7.1.min.js"></script>
</head>
<body onload="init()">
<canvas id="canvas" width="500" height="400" style="border: black solid 1px"></canvas>
```

```
</body>
<script>
    var stage;
    function init() {
        stage = new createjs.Stage(document.getElementById('canvas'));
        createjs.Ticker.addEventListener("tick", handleTick);
        createjs.Ticker.setFPS(60);
        start();
    }
    function handleTick(e) {
        stage.update();
    }
    function start() {
        var circle = new createjs.Shape();
        circle.graphics.beginFill('#0000FF').drawCircle(0, 0, 50);
        circle.x = stage.canvas.width / 2;
        circle.y = stage.canvas.height / 2;
        circle.name = 'Blue Circle';
        stage.addChild(circle);

        circle.addEventListener('dblclick', function (e) {
            alert(e.target.name + '被双击了！');
        });
    }
    function trollAttacked(e){
        alert('ouch');
    }
</script>
</html>
```

✧ 代码解析

createjs.Ticker.addEventListener("tick", handleTick); createjs.Ticker.setFPS(60);以 60 帧/秒的速度调用 handleTick 函数去更新舞台；

start();函数在 canvas 画图区域中间绘制半径为 50 的蓝色圆形；circle.addEventListener('dblclick', function (e) {alert(e.target.name + '被双击了！');});响应鼠标双击事件。

程序运行效果如图 3-28 所示。

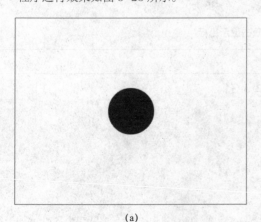
(a)

(b)

图 3-28 鼠标事件 2

【案例 3-17】CreateJS 鼠标移入移出效果

代码清单 3-32

```html
<!DOCTYPE html>
<html>
<head>
<title></title>
<script src="../lib/easeljs-0.7.1.min.js"></script>
</head>
<body onload="init()">
<canvas id="canvas" width="400" height="200" style="border: black solid 1px"></canvas>
</body>
<script>
    var stage;
    var circle;
    function init() {
        stage = new createjs.Stage(document.getElementById('canvas'));
        createjs.Ticker.addEventListener("tick", handleTick);
        createjs.Ticker.setFPS(60);
        start();
    }
    function handleTick(e) {
        stage.update();
    }
    function start() {
        stage.enableMouseOver();
        circle = new createjs.Shape();
        circle.graphics.beginFill('#0000FF').drawCircle(0, 0, 50);
        circle.set({x:stage.canvas.width / 2});
        circle.x = stage.canvas.width / 2;
        circle.y = stage.canvas.height / 2;
        stage.addChild(circle);
        circle.cursor = 'pointer';
        circle.addEventListener("mouseover", function (e) {
            circle.alpha = .5;
        });
        circle.addEventListener("mouseout", function (e) {
            circle.alpha = 1;
        });
    }
</script>
</html>
```

❖ 代码解析

circle.cursor = 'pointer';设置鼠标的形状为箭头形状;

circle.addEventListener("mouseover", function (e) {circle.alpha = .5; });设置鼠标移入效果为圆形半透明;

circle.addEventListener("mouseout", function (e) {circle.alpha = 1; });设置鼠标移出效果为圆形透明度消失;

程序运行效果如图 3-29 所示。

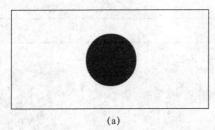
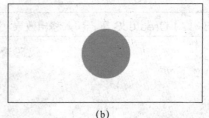

(a)　　　　　　　　　　　　　　(b)

图 3-29　鼠标移入移出效果

【案例 3-18】CreateJS 鼠标拖动效果

代码清单 3-33

```
<!DOCTYPE html>
<html>
<head>
<title></title>
<script src="../lib/easeljs-0.7.1.min.js"></script>
</head>
<body onload="init()">
<canvas id="canvas" width="500" height="300" style="border: black solid 1px"></canvas>
</body>
<script>
    var stage;

    function init(){
        stage = new createjs.Stage(document.getElementById('canvas'));
        createjs.Ticker.addEventListener("tick", handleTick);
        createjs.Ticker.setFPS(60);
        start();
    }
    function handleTick(e){
        stage.update();
    }
    function start(){
        var circle = new createjs.Shape();
        circle.graphics.beginFill('#0000FF').drawCircle(0,0,50);
        circle.x = stage.canvas.width / 2;
        circle.y = stage.canvas.height / 2;

        circle.addEventListener('mousedown',function(e){
            stage.addEventListener('stagemousemove',function(e){
                circle.x = stage.mouseX;
                circle.y = stage.mouseY;
            });
            stage.addEventListener('stagemouseup',function(e){
                e.target.removeAllEventListeners();
            });
        });
        stage.addChild(circle);
    }
</script>
```

</html>

❖ 代码解析

circle.addEventListener('mousedown',function(e){ stage.addEventListener('stagemousemove',function(e){ circle.x = stage.mouseX; circle.y = stage.mouseY; });设置鼠标点中与移动的效果为圆形的位置随着鼠标的（x，y）轴变化而变化；

stage.addEventListener('stagemouseup',function(e){e.target.removeAllEventListeners();});设置鼠标放开效果为清除所有事件。

程序运行结果如图 3-30 所示。

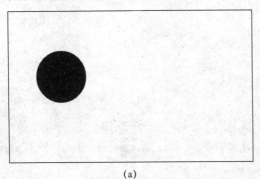 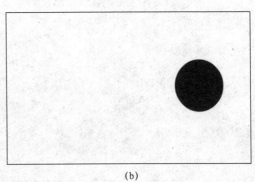

(a)　　　　　　　　　　　　　　　　　　(b)

图 3-30　鼠标移动效果

【案例 3-19】CreateJS 键盘事件

代码清单 3-34

```
<!DOCTYPE html>
<html>
<head>
<title></title>
<script src="../lib/easeljs-0.7.1.min.js"></script>
</head>
<body onload="init()">
<canvas id="canvas" width="1000" height="800" style="border: black solid 1px"></canvas>
</body>
<script>
    var stage;
    function init(){
        stage = new createjs.Stage(document.getElementById('canvas'));
        createjs.Ticker.addEventListener("tick", handleTick);
        createjs.Ticker.setFPS(60);
        start();
    }
    const ARROW_KEY_LEFT = 37;
    const ARROW_KEY_UP = 38;
    const ARROW_KEY_RIGHT = 39;
    const ARROW_KEY_DOWN = 40;
    function start(){
        window.onkeydown = onDPad;
    }
    function onDPad(e){
        switch (e.keyCode){
```

```
                case ARROW_KEY_LEFT:
                    console.log('move left');
                    break;
                case ARROW_KEY_UP:
                    console.log('move up');
                    break;
                case ARROW_KEY_RIGHT:
                    console.log('move right');
                    break;
                case ARROW_KEY_DOWN:
                    console.log('move down');
                    break;
            }
        }
        function handleTick(e){
            stage.update();
        }
    </script>
</html>
```

❖ 代码解析

ARROW_KEY_LEFT = 37；ARROW_KEY_UP = 38；ARROW_KEY_RIGHT = 39；ARROW_KEY_DOWN = 40；设置键盘的左、上、右、下键为37、38、39、40；

start()函数中 window.onkeydown = onDPad；用 onkeydown 函数捕捉键盘按下事件；

onDPad(e)中的 switch 分别对键盘 ARROW_KEY_LEFT、ARROW_KEY_UP、ARROW_KEY_RIGHT、ARROW_KEY_DOWN 进行响应：分别在 console 控制台上显示'move left'、'move up'、'move right'和'move right'信息。

在Google浏览器Throme上点F12调出开发者工具，程序运行效果如图3-31所示，在控制台console中可以看到键盘移动的方位。

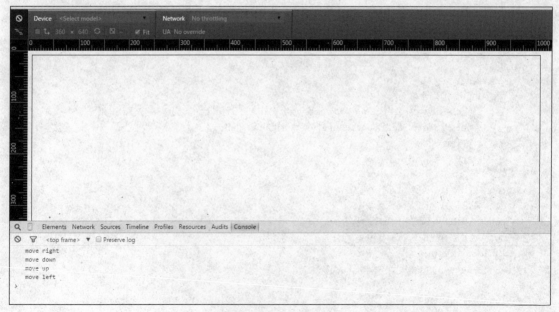

图3-31 键盘事件

【案例 3-20】CreateJS 图形随着键盘的移动而移动

代码清单 3-35

```html
<!DOCTYPE html>
<html>
<head>
<title></title>
<script src="../lib/easeljs-0.7.1.min.js"></script>
</head>
<body onload="init()">
<canvas id="canvas" width="600" height="100" style="border: black solid 1px"></canvas>
</body>
<script>
    const ARROW_KEY_LEFT = 37;
    const ARROW_KEY_RIGHT = 39;
    var stage,padel;
    var leftKeyDown,rightKeyDown = false;
    function init() {
        stage = new createjs.Stage(document.getElementById('canvas'));
        createjs.Ticker.addEventListener("tick", tick);
        createjs.Ticker.setFPS(60);
        start();
    }
    function start() {
        padel = new createjs.Shape();
        padel.graphics.beginFill('#0000FF').drawRect(0, 0, 100, 20);
        padel.width = 100;
        padel.x = padel.nextX = 0;
        padel.y = stage.canvas.height - 20;
        stage.addChild(padel);
        //handle keys
        window.onkeydown = movePadel;
        window.onkeyup = stopPadel;
    }
    function movePadel(e) {
        e = !e ? window.event : e;
        switch (e.keyCode) {
            case ARROW_KEY_LEFT:
                leftKeyDown = true;
                break;
            case ARROW_KEY_RIGHT:
                rightKeyDown = true;
                break;
        }
    }
    function stopPadel(e) {
        e = !e ? window.event : e;
        switch (e.keyCode) {
            case 37:
                leftKeyDown = false;
```

```
                    break;
                case 39:
                    rightKeyDown = false;
                    break;
            }
        }
        function update() {
            var nextX = padel.x;
            if (leftKeyDown) {
                nextX = padel.x - 10;
                if(nextX < 0){
                    nextX = 0;
                }
            }
            else if (rightKeyDown) {
                nextX = padel.x + 10;
                if(nextX > stage.canvas.width - padel.width){
                    nextX = stage.canvas.width - padel.width;
                }
            }
            padel.nextX = nextX;
        }
        function render() {
            padel.x = padel.nextX;
        }
        function tick(e) {
            update();
            render();
            stage.update();
        }
    </script>
</html>
```

❖ 代码解析

window.onkeydown = movePadel; window.onkeyup = stopPadel;对键盘的按下事件和键盘松开事件定义函数响应。

movePadel(e)通过 switch 对键盘左移参数 leftKeyDown、右移参数 rightKeyDown 进行赋值；

stopPadel(e) 通过 switch 对键盘左移参数 leftKeyDown、右移参数 rightKeyDown 进行清空；

function tick(e) 函数包含 update()、render()和 stage.update()更新舞台函数，其中 update()函数规定左移右移的速度为 nextX = padel.x + 10 和 nextX = padel.x - 10，当遇到边界时 if(nextX < 0){ nextX = 0; }和 if(nextX > stage.canvas.width - padel.width){ nextX = stage.canvas.width - padel.width; }}，位置在左边界和右边界停止运动；

render() 函数{ padel.x = padel.nextX; }定义下一次运动位置。

程序运行效果如图 3-32 所示。

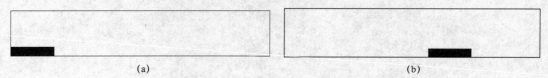

图 3-32 图形随着键盘的移动而移动

3.4 CreateJS 控件

CreateJS 包含非常多好用的控件，这些控件可以减少代码的开发量，本节使用几个案例来分别讲解几个 CreateJS 控件的使用。

3.4.1 Text

在显示列表显示一行或多行动态文本（用户不可编辑）。值得注意的是，作为一种替代文本，可以使用 localToGlobal 方法或使用 DOMElement 在相应的 HTML 文本的上方或下方来添加文本。请注意，Text 不支持 HTML 文本，并且一次只显示一个字体样式。要使用多个字体样式，需要创建多个文本实例，并手动定位它们。

使用方法：

代码清单 3-36

```
var text = new createjs.Text("Hello World", "20px Arial", "#ff7700");
 text.x = 100;
 text.textBaseline = "alphabetic";
```

使用 Text 可以绘制文字，包括文字的颜色大小等都可以随意变更。

极客学院在线视频学习网址：
http://www.jikexueyuan.com/course/277_1.html
手机扫描二维码

CreateJS 控件-Text

【案例 3-21】CreateJS 文字效果

代码清单 3-37

```
<!DOCTYPE html>
<html>
<head>
<title></title>
<script src="../lib/easeljs-0.7.1.min.js"></script>
</head>
<body onload="init()">
<canvas id="canvas" width="500" height="300" style="border: black solid 1px"></canvas>
</body>
<script>
    var stage;
    function init(){
        stage = new createjs.Stage(document.getElementById('canvas'));
        createjs.Ticker.addEventListener("tick", handleTick);
        createjs.Ticker.setFPS(60);
```

```
            start();
        }
        function start(){
            var text = new createjs.Text("游戏结束", "20px Arial", "#ff7700");
            text.textBaseline = "middle";
            text.textAlign = "center";
            text.x = stage.canvas.width / 2;
            text.y = stage.canvas.height / 2;
            stage.addChild(text);
        }
        function handleTick(e){
            stage.update();
        }
</script>
</html>
```

✧ 代码解析

var text = new createjs.Text("游戏结束", "20px Arial", "#ff7700");定义文字的内容、格式和颜色；

text.textBaseline = "middle";定义文字的下基线对齐方式；

text.textAlign = "center";定义文字的水平对齐方式；

text.x = stage.canvas.width / 2; text.y = stage.canvas.height / 2;定义文字的位置为舞台正中间。

程序运行效果如图 3-33 所示。

图 3-33　文本效果

【案例 3-22】CreateJS 数字自增效果

代码清单 3-38

```
var rect;
var count = 0 ;
canvas = document.getElementById("gamView");
stage = new createjs.Stage(canvas);
text = new createjs.Text("test on the Canvas ... 0!", "36px Arial", "#FFF");
text.x = 10;
text.y = 10;
text.rotation   = 20;
```

```
stage.addChild(text);
rect = new createjs.Shape();
rect.x = text.x;
rect.y = text.y;
rect.rotation = text.rotation;
stage.addChildAt(rect,0);
createjs.Ticker.setFPS(100);
createjs.Ticker.addEventListener("tick", handlertick);
function handlertick(e){
    count++;
    text.text = "test on the Canvas ..." + count +"!";
    rect.graphics.clear().beginFill("#f00").drawRect(-10,-10,text.getMeasuredWidth()+20,50);
    stage.update(e);
}
```

❖ 代码解析

text = new createjs.Text("test on the Canvas ... 0!", "36px Arial", "#FFF");定义一个默认为"test on the Canvas ... 0!"的文字效果；

text.x = 10;text.y = 10;text.rotation = 20;定义文字的 x、y 轴和旋转角度；

rect 在舞台上添加一个与 text 同样旋转角度的矩形，与 text 形成重叠效果；

createjs.Ticker.setFPS(100);createjs.Ticker.addEventListener("tick", handlertick);以 100 帧/秒的速度运行 handlertick 函数；

handlertick(e)函数实现 text 内容中数值的递增，其中 rect.graphics.clear().beginFill("#f00").drawRect(-10,-10,text.getMeasuredWidth()+20,50);实现清除重画宽度为 text.getMeasuredWidth()+20 的矩形效果，以避免由于数值过大而被遮住。

程序运行效果如图 3-34 所示。

图 3-34　数字自增效果

3.4.2 BitMap

控件 BitMap 的应用，掌握了 BitMap 之后可以快速地对图片做处理。

极客学院在线视频学习网址：
http://www.jikexueyuan.com/course/277_2.html
手机扫描二维码

CreateJS 控件-BitMap

【案例 3-23】使用 CreateJS BitMap 实现图片加载效果

代码清单 3-39

```
var stage = new createjs.Stage("gamView");
var gameView = new createjs.Container();
stage.addChild(gameView);
var bitmap = new createjs.Bitmap("9.jpg");
gameView.addChild(bitmap);

createjs.Ticker.setFPS(30);
createjs.Ticker.addEventListener("tick", function(){
    stage.update();
});
```

◆ 代码解析

var bitmap = new createjs.Bitmap("9.jpg");gameView.addChild(bitmap);加载图片，并将图片放置于画布中。

程序运行效果如图 3-35 所示。

图 3-35　CreateJS 位图加载

【案例 3-24】使用 CreateJS BitMap 实现图片加载效果 2

代码清单 3-40

```
<!DOCTYPE html>
<html>
```

```
<head>
<title></title>
<script src="../lib/easeljs-0.7.1.min.js"></script>
</head>
<body onload="init()">
<canvas id="canvas" width="300" height="240" style="border: black solid 1px"></canvas>
</body>
<script>
    var stage, img;
    function init() {
        stage = new createjs.Stage(document.getElementById('canvas'));
        createjs.Ticker.addEventListener("tick", handleTick);
        createjs.Ticker.setFPS(60);
        img = new Image();
        img.addEventListener('load', drawFrank);
        img.src = 'img/frank.png';
    }
    function drawFrank(e) {
        var frank = new createjs.Bitmap(e.target);
        stage.addChild(frank);
        stage.update();
    }
    function handleTick() {
        stage.update();
    }
</script>
</html>
```

❖ 代码解析

img = new Image();img.addEventListener('load', drawFrank); img.src = 'img/frank.png';定义图形对象，在图形对象的加载事件中调用 drawFrank 函数；

drawFrank(e) 函数中 var frank = new createjs.Bitmap(e.target); stage.addChild(frank); 用 Bitmap 将定义的图形对象载入，最后加载到舞台上。

程序运行效果如图 3-36 所示。

图 3-36 位图加载效果 2

3.4.3 MovieClip

掌握了 MovieClip 可以更好地处理动画。movieclip-0.7.1.min.js 在下载的代码包 EaselJS-0.8.1\lib 中。

极客学院在线视频学习网址：
http://www.jikexueyuan.com/course/277_3.html
手机扫描二维码

CreateJS 控件-MovieClip

【案例 3-25】使用 MovieClip 实现小球移动效果

代码清单 3-41

```
<!DOCTYPE html>
<html>
<head lang="en">
<meta charset="UTF-8">
<title></title>
<script src="easeljs-0.7.1.min.js"></script>
<script src="tweenjs-0.5.1.min.js"></script>
<script src="movieclip-0.8.1.min.js"></script>
</head>
<body>
<canvas id="gamView" width="400px" height="400px"
style="background-color:#ffffff"></canvas>
<script src="app.js"></script>
</body>
</html>
var stage = new createjs.Stage("gamView");
createjs.Ticker.setFPS(30);
createjs.Ticker.addEventListener("tick", stage);
var mc = new createjs.MovieClip(null,0,true,{ start:50,stop:0});
stage.addChild(mc);
var state1 = new createjs.Shape(new
createjs.Graphics().beginFill("#999999").drawCircle(0,100,30));
var state2 = new createjs.Shape(new
createjs.Graphics().beginFill("#555555").drawCircle(0,100,30));
mc.timeline.addTween(createjs.Tween.get(state1).to({x:0}).to({x:400},100).to({x:0},100));
mc.timeline.addTween(createjs.Tween.get(state2).to({x:400}).to({x:0},100).to({x:400},100));
mc.gotoAndPlay("start");
//mc.gotoAndPlay("stop");
```

✧ 代码解析

\<script src="easeljs-0.7.1.min.js"\>\</script\>\<script src="tweenjs-0.5.1.min.js"\>\</script\>\<script

src="movieclip-0.7.1.min.js"></script>引入 easeljs-0.7.1.min.js、tweenjs-0.5.1.min.js 和 movieclip-0.7.1.min.js 包；

var mc = new createjs.MovieClip(null,0,true,{ start:50,stop:0});创建一个 MovieClip 对象，其中第一个参数为渲染（null 表示本案例没有渲染），第二参数为起始位置为 0，第三个参数为是否可循环执行（true 为可循环执行，false 为不可循环执行），第四个参数为时间线（start:50 表示当前线为 50，stop:0 表示停止线为 0）；

stage.addChild(mc);把 mc 添加到舞台上；

state1、state2 绘制两个不同颜色的圆形；

mc.timeline.addTween(createjs.Tween.get(state1).to({x:0}).to({x:400},100).to({x:0},100));设置 mc 的时间线（图形运行的动画），制定第一个图形 state1 的 to({x:0}).to({x:400},100) x 坐标从 0 开始到 400 结束，时间线为 100，而 to({x:0},100)说明 x 轴重新回到 0，时间线还是为 100；

mc.timeline.addTween(createjs.Tween.get(state2).to({x:400}).to({x:0},100).to({x:400},100));设置 mc 的时间线（图形运行的动画），制定第二个图形 state2 的 to({x:400}).to({x: 0},100) x 坐标从 400 开始到 0 结束，时间线为 100，而 to({x:400},100)说明 x 轴重新回到 400，时间线还是为 100；

mc.gotoAndPlay("start");设置图形从 start 位置 50 开始运行。

程序运行效果如图 3-37 所示。

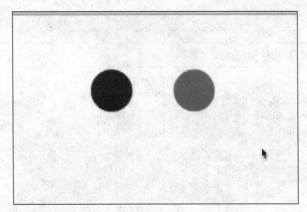

图 3-37　movieClip 实现小球从中心移动

mc.gotoAndPlay("stop");设置图形从 stop 位置 0 开始运行。

程序运行效果如图 3-38 所示。

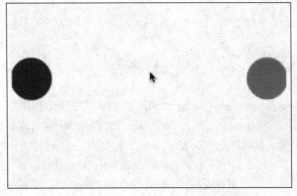

图 3-38　movieClip 实现小球从边缘移动

3.4.4 Sprite

从 spritesheet 显示一个帧或帧序列（即动画）实例。精灵片是一系列的图像（通常是动画帧）合并成一个单一的图像。例如，一个动画由 8100×100 图像可以组合成一个 400×200 精灵表。可以显示单个帧，播放帧作为一个动画，甚至序列动画一起播放。

使用语法：

代码清单 3-42

```
var instance = new createjs.Sprite(spriteSheet);
instance.gotoAndStop("frameName");
```

极客学院在线视频学习网址：
http://www.jikexueyuan.com/course/277_4.html
手机扫描二维码

CreateJS 控件-Sprite

【案例 3-26】使用 Sprite 实现跑跳效果

代码清单 3-43

```
var stage = new createjs.Stage("gamView");
var ss = new createjs.SpriteSheet({
    "images":["runningGrant.png"],
    "frames":{
        "height":292,
        "width":165,
        "count":64},
    "animations":{
        "run":[0,25,"jump"],
        "jump":[26,63,"run"]}
});
var s = new createjs.Sprite(ss,"run");
s.x = 100;
s.y = 100;
stage.addChild(s);
createjs.Ticker.setFPS(60);
createjs.Ticker.addEventListener("tick",stage);
```

✧ 代码解析

var ss = new createjs.SpriteSheet()定义一个图像序列，其中"images":["runningGrant.png"]，定义图像的源文件路径；"frames":{"height":292, "width":165, "count":64}，定义剪切图片的高宽和个数；"animations":定义动画效果，其中"run":[0,25,"jump"]，定义跑的效果为从 0 到 25 图片，跑运行完继续运行"jump"的效果，"jump":[26,63,"run"]}定义跳的效果为从 26 到 63 图片，跳运行完继续运行"run"的效果，这样可以实现 run、jump 的循环执行；

var s = new createjs.Sprite(ss,"run");s.x = 100;s.y = 100;定义 sprite 对象并将 ss 加载进去,设置其 x、y 轴坐标;

createjs.Ticker.setFPS(60);createjs.Ticker.addEventListener("tick",stage);实现 60 帧/秒的跑跳效果。

程序运行效果如图 3-39 所示。

(a)　　　　　　　　　　　　　　　　(b)

图 3-39　sprite 实现跑跳效果

【案例 3-27】使用 Sprite 实现小猪飞翔效果

代码清单 3-44

```
<!DOCTYPE html>
<html>
<head>
<title></title>
<script src="../lib/easeljs-0.7.1.min.js"></script>
</head>
<body onload="init()">
<canvas id="canvas" width="1000" height="800" style="border: black solid 1px;background-color: #add8e6"></canvas>
</body>
<script>
    var stage;
    function init() {
        stage = new createjs.Stage(document.getElementById('canvas'));
        createjs.Ticker.addEventListener("tick", runGame);
        createjs.Ticker.setFPS(60);
        addPig();
    }
    function addPig() {
        var ss = new createjs.SpriteSheet(data);
        pig = new createjs.Sprite(ss, 'all');
        stage.addChild(pig);
    }
    function runGame(e) {
        pig.y += 1;
        pig.x += 1;
        stage.update();
```

```
                if (pig.y > stage.canvas.height) {
                    pig.x = pig.y = 0;
                }
            }
        var data = {
            "images":["pig_0.png", "pig_1.png"],
            "frames":[
                [2, 2, 252, 252, 0, -61, -127],
....//查阅源代码
                [1538, 1794, 252, 252, 1, -61, -127]
            ],
            "animations":{
                "all":{
                    "frames":[0, 1, 2, 3, 4, 5, 6, 7, 8, 9, 10, 11, 12, 13, 14, 15, 16, 17, 18, 19, 20, 21,
22, 23, 24, 25, 26, 27, 28, 29, 30, 31, 32, 33, 34, 35, 36, 37, 38, 39, 40, 41, 42, 43, 44, 45, 46, 47, 48, 49,
50, 51, 52, 53, 54, 55, 56, 57, 58, 59, 60, 61, 62, 63, 64, 65, 66, 67, 68, 69, 70, 71, 72, 73, 74, 75, 76, 77,
78, 79, 80, 81, 82, 83, 84, 85, 86, 87, 88, 89, 90, 91, 92, 93, 94, 95, 96, 97, 98, 99, 100, 101, 102, 103,
104, 105, 106, 107, 108, 109, 110, 111, 112, 113, 114, 115, 116, 117, 118, 119, 120, 121, 122, 123, 124,
125, 126],
                    "speed":.4
                }
            }
        };
</script>
</html>
```

❖ 代码解析

<body onload="init()">页面加载时运行 init()函数；

init()函数 createjs.Ticker.addEventListener("tick", runGame); createjs.Ticker.setFPS(60);以 60帧/秒的速度运行 runGame 函数；之后再运行 addPig()函数；

runGame(e) 函数设置小猪飞行的轨迹，速度为 x/y 轴+1 当 (pig.y > stage.canvas.height) 小猪 y 轴大于舞台高度时，重新回到（0,0）的位置；

addPig() 函数 ss = new createjs.SpriteSheet(data); pig = new createjs.Sprite(ss, 'all');用 sprite()函数设置小猪的数据源与移动信息。

程序运行效果如图 3-40 所示。

图 3-40 小猪飞翔效果

3.4.5 DOMElement

DOMElement 通过 CreateJS 可以直接操作元素。DOMElement 让您联想到 DOMElement HtmlElement 显示列表，但是它是在 Container 添加的，却被转化为 DOM 子元素。DOMElement 不是渲染画布，所以保留了任何相对于 Canvas 的 z-index 属性（被绘制在 Canvas 的前面或后面）。一个 DOMElement 位置相对于父节点的 DOM，这 DOM 对象被添加到一个 div，DOMElement 用于定位 HTML 元素之上的画布的内容和想显示在画布上的边界元素。例如，一个具有丰富的 HTML 内容提示和鼠标交互等。DOMElement t 实例是 HTML 对象，所以不参与 easeljs 鼠标事件。从 DOMElement 获取鼠标事件，必须添加处理程序的 HtmlElement 方法。

语法格式如下：

代码清单 3-45

```
var domElement = new createjs.DOMElement(htmlElement);
 domElement.htmlElement.onclick = function() {
console.log("clicked");
 }
```

极客学院在线视频学习网址：
http://www.jikexueyuan.com/course/277_5.html
手机扫描二维码

CreateJS 控件-DOMElement

【案例 3-28】使用 DOMElement 实现容器中加载 HTML 元素

代码清单 3-46

```
<!DOCTYPE html>
<html>
<head lang="en">
<meta charset="UTF-8">
<title></title>
<script src="easeljs-0.7.1.min.js"></script>
</head>
<body>
<div id="div" style="background-color: aqua; width: 400px; height: 400px" >
<button id="btn"   width="100px", height="50px" onclick="alert('hello')">按钮</button>
</div>
<canvas id="gamView" width="400px" height="400px"
style="background-color:#ffffff"></canvas>
<script src="app.js"></script>
</body>
</html>
 var stage,container,canvas;
```

```
canvas = document.getElementById("gamView");
stage = new createjs.Stage(canvas);
container = new createjs.Container();
stage.addChild(container);
container.x = 100;
container.y = 100;
var content = new createjs.DOMElement("div");
container.addChild(content);
stage.update();
```

✧ 代码解析

<div id="div" style="background-color: aqua; width: 400px; height: 400px" > <button id="btn" width="100px", height="50px" onclick="alert('hello')">按钮</button> </div>定义 div 对象属性；

var content = new createjs.DOMElement("div"); container.addChild(content);将 div 元素转换为 createjs 对象，并把这个对象添加到 container 容器中。

container.x = 100;container.y = 100;container 容器的 x、y 轴发生变化，可以从程序运行效果中看到 div 按钮的位置也跟着变化了。

程序运行效果如图 3-41 所示。

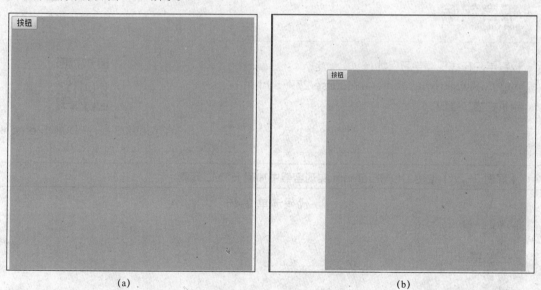

图 3-41　DOMElement 效果图 1

【案例 3-29】使用 DOMElement 实现 HTML 元素动画

代码清单 3-47

```
<!DOCTYPE html>
<html xmlns="http://www.w3.org/1999/html">
<head>
<title></title>
<script src="../lib/easeljs-0.7.1.min.js"></script>
<script src="../lib/tweenjs-0.5.1.min.js"></script>
</head>
<body onload="init()">
```

```html
<div id="gameHolder">
    <div id="instructions" style="width: 400px;height: 300px;border: dashed 2 #008b8b;text-align: center;visibility: hidden">
        <h3 style="font-family:arial;">游戏说明</h3>
        <p><strong>点击</strong><span style="color:red">红色</span>按钮。</p>
        <p>Make sure you click them <span style="text-decoration: underline">all</span> before time runs out!</p>
        <p>Rack up <i>as many points</i> as you can to reach the <span style="color:blue">BLUE</span> level.</p>
        <h2 style="font-weight: bold;margin-top:30px">GOOD LUCK!</h2>
    </div>
    <canvas id="canvas" width="500" height="400" style="border: black solid 1px"></canvas>
</div>
</body>
<script>
    var stage;
    function init() {
        stage = new createjs.Stage(document.getElementById('canvas'));
        createjs.Ticker.addEventListener("tick", handleTick);
        createjs.Ticker.setFPS(60);
        start();
    }
    function start() {
        var de = new createjs.DOMElement(instructions);
        de.alpha = 0;
        de.regX = 200;
        de.x = stage.canvas.width / 2;
        de.y = 0;
        stage.addChild(de);
        createjs.Tween.get(de).wait(1000).to({y:40, alpha:1}, 2000, createjs.Ease.quadOut);
    }
    function handleTick(e) {
        stage.update();
    }
</script>
</html>
```

❖ 代码解析

\<div id="instructions" ………\</div>定义 id 为 instructions 的 div 元素;

\<body onload="init()">程序加载默认执行 init()函数;

init() 函数定义舞台和更新频率,调用 start()函数 var de = new createjs.DOMElement(instructions);在舞台上创建 instructions 的元素;

de.alpha = 0; de.regX = 200; de.x = stage.canvas.width / 2; de.y = 0; stage.addChild(de);定义 de 元素的透明度、宽度、x/y,并将其加载到舞台上;

createjs.Tween.get(de).wait(1000).to({y:40, alpha:1}, 2000, createjs.Ease.quadOut);设置渐进渐出的动画效果。

程序运行效果如图 3-42 所示。

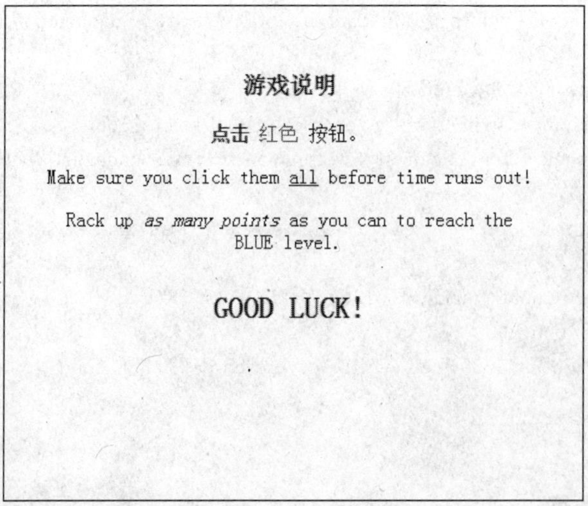

图 3-42 DOMElement 动画效果图

3.5 Tween 函数包

CreateJS 还有很多保证好的功能，本节掌握 TweenJS 的 CSSPlugin、Ease、MotionGuidePlugin、TweenJS 的 Tween。

3.5.1 CSSPlugin

本节讲解 TweenJS 的 CSSPlugin，了解在 TweenJS 中如何操作 CSS。其源代码在下载的源代码包：TweenJS-0.6.1\src\tweenjs 下。

极客学院
jikexueyuan.com

极客学院在线视频学习网址：
http://www.jikexueyuan.com/course/278_1.html
手机扫描二维码

CreateJS TweenJS-CSSPlugin

【案例 3-30】使用 CSSPlugin 包创建 Div 元素

了解 TweenJS 的 CSSPlugin，以及在 TweenJS 中如何操作 CSS。

代码清单 3-48

```
<!DOCTYPE html>
<html>
<head lang="en">
<meta charset="UTF-8">
<title></title>
```

```
<script src="easeljs-0.7.1.min.js"></script>
<script src="tweenjs-0.6.1.min.js"></script>
<script src="CSSPlugin.js"></script>
</head>
<body>
<script src="app.js"></script>
</body>
</html>
createjs.CSSPlugin.install(createjs.Tween);
var box = document.createElement("div");
box.style.width = "400px";
box.style.height = "400px";
box.style.position = "absolute";
box.style.backgroundColor= "#ff0000";
document.body.appendChild(box);
```

◆ 代码解析

\<script src="easeljs-0.7.1.min.js"\>\</script\>\<script src="tweenjs-0.6.1.min.js"\>\</script\>\<script src="CSSPlugin.js"\>\</script\>引入 easeljs-0.7.1.min.js、tweenjs-0.6.1.min.js 和 tweenjs-0.6.1.min.js 函数包；

createjs.CSSPlugin.install(createjs.Tween)；将 createjs.Tween 导入获取 CSSPlugin 对象；

var box = document.createElement("div")；通过 box 来获取 Div 元素；

box.style.width = "400px"；box.style.height = "400px"；box.style.position = "absolute"；box.style.backgroundColor= "#ff0000"；设置 box 宽度、高度和绝对位置、背景颜色为红色；

document.body.appendChild(box)；将 box 对象加入 document 的 body 中去，实现用 createjs 操作 CSS 对象的效果。

程序运行结果如图 3-43 所示。

图 3-43 使用 CSSPlugin 包创建 Div 元素

3.5.2 Ease

本节讲解 TweenJS 的 Ease，Ease 包含很多动画曲线，可以根据自己的需求添加。源代码在 TweenJS-0.6.1\src\tweenjs 中。

极客学院在线视频学习网址：
http://www.jikexueyuan.com/course/278_2.html
手机扫描二维码

CreateJS TweenJS-Ease

【案例 3-31】使用 Ease 包创建图形效果

代码清单 3-49

```
var stage = new createjs.Stage("gameView");
var circle = new createjs.Shape();
circle.graphics.beginFill("#FF0000").drawCircle(50,50,50);
stage.addChild(circle);
createjs.Tween.get(circle,{loop:false},true)
    .to({x:500,y:0,alpha:0.1},1000,createjs.Ease.backIn)
createjs.Ticker.setFPS(30);
createjs.Ticker.addEventListener("tick", stage);
```

✧ 代码解析

createjs.Tween.get(circle,{loop:false},true)操作对象为 circle，{loop:false}不重复执行；
.to({x:500,y:0,alpha:0.1},1000,createjs.Ease.backIn 从开始位置移动到（500,0），并将透明度变为 0.1，效果为 createjs.Ease.backIn；

程序运行效果如图 3-44 所示。

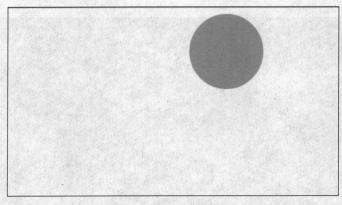

图 3-44 ease 动画效果 1

将代码修改为如下：

代码清单 3-50

```
createjs.Tween.get(circle,{loop:false},true)
.to({x:500,y:0,alpha:0.1},1000,createjs.Ease.elasticInOut)
    .to({x:0},1000,createjs.Ease.backIn)
    .wait(1000)
    .to({alpha:1},100);
```

```
createjs.Ticker.setFPS(30);
createjs.Ticker.addEventListener("tick", stage);
```

❖ 代码解析

to({x:500,y:0,alpha:0.1},1000,createjs.Ease.elasticInOut) 从开始位置移动到（500,0），并将透明度变为 0.1，效果为 createjs.Ease.elasticInOut；

.to({x:0},1000,createjs.Ease.backIn) 从(500,0)位置移动到(0,0)，效果为 elasticInOut；

.wait(1000) .to({alpha:1},100)；等待 1000 毫秒，将透明度变为 1；

程序运行效果如图 3-45 所示（更多动画效果请运行代码）。

图 3-45　ease 动画效果 2

TweenJS 的主页上有所有图形的移动效果，如图 3-46 所示，大家可以参照一下。

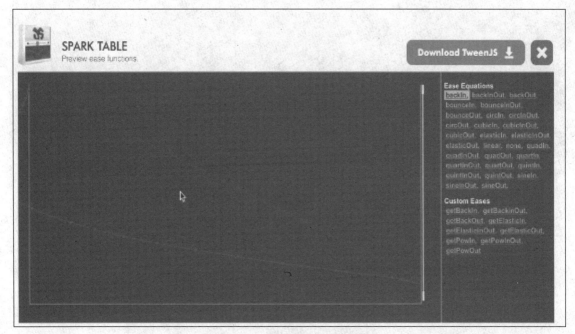

图 3-46　各类图形移动效果图

3.5.3　MotionGuidePlugin

本节讲解 TweenJS 的 MotionGuidePlugin，程序源代码在 TweenJS-0.6.1\src\tweenjs 下。

极客学院在线视频学习网址：
http://www.jikexueyuan.com/course/278_3.html
手机扫描二维码

CreateJS TweenJS-MotionGuidePlugin

【案例 3-32】使用 MotionGuidePlugin 包实现图形变形效果

代码清单 3-51

```
<!DOCTYPE html>
<html>
<head lang="en">
<meta charset="UTF-8">
<title></title>
<script src="easeljs-0.7.1.min.js"></script>
<script src="tweenjs-0.6.1.min.js"></script>
<script src="MotionGuidePlugin.js"></script>
</head>
<body>
<canvas id="gameView" width="400px" height="400px" ></canvas>
<script src="app.js"></script>
</body>
</html>
var canvas;
var stage;
createjs.MotionGuidePlugin.install(createjs.Tween);
var stage = new createjs.Stage("gameView");
var ball = new createjs.Shape();
var b = ball.graphics;
b.beginFill("#FF0000").drawCircle(0,0,50);
b.endFill();
createjs.Tween.get(ball,{loop:false},true)
    .to({guide:{path:[100,100,400,100,200,300],orient:true}},5000);
stage.addChild(ball);
createjs.Ticker.setFPS(30);
createjs.Ticker.addEventListener("tick", stage);
```

❖ 代码解析

createjs.MotionGuidePlugin.install(createjs.Tween); 导入 createjs.Tween 创建 MotionGuidePlugin 实体；

createjs.Tween.get(ball,{loop:false},true) 操作对象为 ball，{loop:false}不重复执行；

.to({guide:{path:[100,100,400,100,200,300],orient:true}},5000);运行的轨迹为 guide 数据制定的路径（100,100）→（400,100）→（200,300），运行周期为 5000 毫秒。

程序运行效果如图 3-47 所示（更多动画效果请运行代码）。

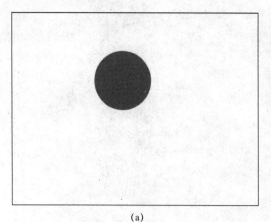
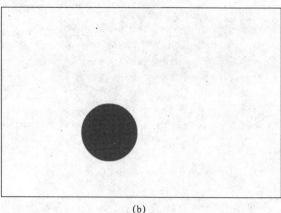

(a)　　　　　　　　　　　　　　　(b)

图 3-47　MotionGuidePlugin 包实现图形变形效果

3.5.4　Tween

本节讲解 Tween 的整体应用效果。Tween 相关的函数包源代码在 Tween 包中。

极客学院在线视频学习网址：
http://www.jikexueyuan.com/course/278_4.html
手机扫描二维码

CreateJSTweenJS-Tween

【案例 3-33】使用 Tween 包实现图形变形效果

代码清单 3-52

```
var canvas;
var stage;
var tweens;
var circleCount = 25;
var activeCount;
canvas = document.getElementById("gameView");
stage = new createjs.Stage(canvas);
tweens = [];
for(var i = 0; i<circleCount; i++){
    var circle = new createjs.Shape();
    circle.graphics.setStrokeStyle(15);
    circle.graphics.beginStroke("#113355");
    circle.graphics.drawCircle(0,0,(i+1)*4);
    circle.compositeOperation = "lighter";
    var   tween = createjs.Tween.get(circle)
        .to({x:300,y:200},(0.5+i*0.04)*1500, createjs.Ease.backInOut.call(tweenComplete));
    tweens.push({tween:tween,ref:circle});
    stage.addChild(circle);
```

```
}
activeCount = circleCount;
stage.addEventListener("stagemouseup",handlerMouseup);
createjs.Ticker.addEventListener("tick",stage);
function handlerMouseup(e){
    for(var i = 0; i<circleCount; i++){
        var ref = tweens[i].ref;
        createjs.Tween.get(ref,{override:true}).to({x:stage.mouseX,y:stage.mouseY},
            (0.5+i*0.04)*1500,createjs.Ease.bounceOut).call(tweenComplete);
    }
    activeCount = circleCount;
}
function tweenComplete(){
    activeCount--;
}
```

❖ 代码解析

for(var i = 0; i<circleCount; i++)用 for 循环实现 25 个圆形效果;

circle.graphics.setStrokeStyle(15); 设置画圆的线型;

circle.graphics.beginStroke("#f0f");circle.graphics.drawCircle(0,0,(i+1)*4);设置颜色和各个圆的初始位置; circle.compositeOperation = "lighter";设置亮度变亮效果;

var tween = createjs.Tween.get(circle)..to({x:300,y:200},(0.5+i*0.04)*1500,createjs.Ease.backInOut.call(tweenComplete)); 创建每个圆形动画从 (300,200) 移动到(0.5+i*0.04)*1500,动画的效果为 backInOut;

tweens.push({tween:tween,ref:circle});把效果放入 tweens 对象中;

stage.addEventListener("stagemouseup",handlerMouseup);创建鼠标 stagemouseup 响应事件;

handlerMouseup(e)函数响应的事件为：用 for(var i = 0; i<circleCount; i++)循环获取 var ref = tweens[i].ref; createjs.Tween.get(ref,{override:true}).to({x:stage.mouseX,y:stage.mouseY}, (0.5+i*0.04)*1500,createjs.Ease.bounceOut).call(tweenComplete);设置每个圆形动画从当前位置移动到鼠标点击的位置(0.5+i*0.04)*1500,动画的效果为 bounceOut;

程序运行效果如图 3-48 所示（更多动画效果请运行代码）。

图 3-48　Tween 动画效果图

案例实战篇

第4章

简单效果案例

■ 前面几章已经掌握了Canvas与CreateJS的基础知识,现在是CreateJS发挥其价值的最佳时机:降低 HTML5 项目的开发难度和开发成本,让开发者以简单、熟悉的方式打造更具现代感的网络交互体验。本章通过帧动画效果、跳舞蝴蝶效果、颜色拼图游戏、图像处理效果、处理跑跳效果及炫酷效果实现等6个综合案例,详细讲解如何在实际应用中配合使用 CreateJS、Canvas 与 CreateJS 来进行开发。

第4章 简单效果案例

4.1 帧动画效果

1. 构建基本窗口，如代码清单 4-1 所示。

代码清单 4-1

```html
<!DOCTYPE html>
<meta charset="utf-8" />
<style type="text/css">
body{text-align:center;}
    #c1{border:1px dotted black}
</style>
<body>
<h2>超级玛丽动画效果</h2>
<img id="img1" src="images/image.png" />
<input id="btnGO" type="button" value="开始" /><br>
<canvas id="c1" width="320" height="200" ></canvas><br>
</body>
</html>
```

2. 构建超级玛丽动画效果，如代码清单 4-2 所示。

代码清单 4-2

```javascript
<script>
var isAnimStart = false; //是否开始动画
var MarioMovie = null; //动画函数
var framen = 0;    //图片切割个数
var frames = []; //保存每帧动画起始坐标,本例图片共有
for (framen=0; framen<15; framen++ ) {
        frames[framen] = [32*framen, 0];
    }
    //定义每帧图像的宽度和高度
var fWidth = 32,
fHeight = 32;
function $(id)
    {
returndocument.getElementById(id);
    }
functioninit()
    {
        //响应onclick事件
        $("btnGO").onclick=function()
    {
    //如果没开始动画，则开始动画
if(!isAnimStart)
        {
        var ctx = $("c1").getContext("2d");
        var fIndex = 0;
        var cX = 160,
            cY = 100;
        animHandle = setInterval(function(){
            ctx.clearRect(0,0,320,200);
ctx.drawImage(img1,
           frames[fIndex][0],frames[fIndex][1],fWidth,fHeight,
```

```
                    cX-64,cY-64,fWidth*4,fHeight*4);
                fIndex++;
                if(fIndex>=frames.length)
                 {
                 fIndex = 0;
                 }
            },100)
                $("btnGO").value = "停止";
        isAnimStart = true;
        }
        else
        {
           $("btnGO").value = "开始";
        clearInterval(animHandle);
        isAnimStart = false;
        }
    }
   }
init();
</script>
```

◆ 代码解析

本案例实现了超级玛丽行走蹲下等的效果，主要原理采用一秒钟连续放映 20 张静态图片的方式形成了动态效果。本案例中主要用 drawImage()函数实现画图效果，用 setInterval()函数实现循环播放，用 clearInterval()函数实现动画的停止。

for (framen=0; framen<15; framen++) {frames[framen] = [32*framen, 0];}将 image.png 图片进行坐标切割，形成 15 个不同的超级玛丽状态图，并把每个图片的横纵坐标放入 frames[]参数中；

ctx.clearRect(0,0,320,200);ctx.drawImage(img1,frames[fIndex][0],frames[fIndex][1],fWidth,fHeight, cX-64,cY-64,fWidth*4,fHeight*4);清空画布后，把当前序列号为 Index 的图片画到(cX-64,cY-64)的位置上，且 fWidth*4,fHeight*4)表示高度和宽度放大四倍。

fIndex++;if(fIndex>=frames.length){ fIndex = 0;} 设置 15 个图像都循环显示完成之后，又从第一个图像开始循环显示；

MarioMovie = setInterval(function(){},100)setInterval 函数让函数体 function 里面的代码以 100 毫秒的速度周期执行，可以调整毫秒值来使帧速度变快或者变慢；

clearInterval(MarioMovie);clearInterval()函数停止 MarioMovie 的动作循环效果。

程序运行效果如图 4-1 所示。

图 4-1　帧动画效果图

4.2 跳舞蝴蝶效果

1. 构建基本窗口，如代码清单 4-3 所示。

代码清单 4-3

```html
<!DOCTYPE html>
<html>
<head>
<title></title>
<script src="../lib/easeljs-0.7.1.min.js"></script>
<script src="../lib/tweenjs-0.5.1.min.js"></script>
<script src="../lib/soundjs-0.5.2.min.js"></script>
<script src="../lib/preloadjs-0.4.1.min.js"></script>
</head>
<body onload="init()">
<canvas id="canvas" width="1000" height="800" style="border: black solid 1px"></canvas>
</body>
```

2. 蝴蝶跳舞效果，如代码清单 4-4 所示。

代码清单 4-4

```
<script>
var stage;
var queue;
function init() {
queue = new createjs.LoadQueue();
queue.installPlugin(createjs.Sound);
queue.addEventListener("complete", loadComplete);
queue.loadManifest([
            {id:"butterfly", src:"images/butterfly.png"},
            {id:"woosh", src:"sounds/woosh.mp3"},
            {id:"chime", src:"sounds/chime.mp3"}
        ]);
    }
function loadComplete() {
setupStage();
buildButterflies();
    }
function setupStage() {
stage = new createjs.Stage(document.getElementById('canvas'));
createjs.Ticker.setFPS(60);
createjs.Ticker.addEventListener("tick", function(){
stage.update();
        });
    }
function buildButterflies() {
var img = queue.getResult("butterfly");
var i, sound, butterfly;
for (i = 0; i < 3; i++) {
butterfly = new createjs.Bitmap(img);
            butterfly.x = i * 200;
```

```
stage.addChild(butterfly);
createjs.Tween.get(butterfly).wait(i * 1000).to({y:100}, 1000, createjs.Ease.quadOut).call(butterflyComplete);
sound = createjs.Sound.play('woosh',createjs.Sound.INTERRUPT_NONE,i * 1000);
            }
        }
function butterflyComplete(){
stage.removeChild(this);
if(!stage.getNumChildren()){
createjs.Sound.play('chime');
            }
        }
function handleTick(e) {
stage.update();
        }
</script>
</html>
```

◆ 代码解析

queue = new createjs.LoadQueue();利用 createjs LoadQueue()资源队列载入的方法来创建一个 queue 对象；

queue.installPlugin(createjs.Sound);载入声音；

queue.addEventListener("complete", loadComplete);载入 loadComplete 事件；

queue.loadManifest([{id:"butterfly", src:"images/butterfly.png"},{id:"woosh", src:"sounds/woosh.mp3"},{id:"chime", src:"sounds/chime.mp3"}]);载入图片和声音资源；

loadComplete()函数中运行 setupStage()和 buildButterflies()函数，其中 setupStage()设置舞台基本属性；buildButterflies()设置三个蝴蝶并排排列效果，createjs.Tween.get(butterfly).wait(i * 1000).to({y:100}, 1000, createjs.Ease.quadOut).call(butterflyComplete);设置从左到右蝴蝶的移动效果，事件延迟 1000×i 毫秒，移动的效果是从初始位置到(x,100)的位置，效果为 quadOut，移动好之后调用 butterflyComplete 函数；

sound = createjs.Sound.play('woosh',createjs.Sound.INTERRUPT_NONE,i * 1000);设置移动过程调用'woosh'的声音效果；

butterflyComplete()函数的 if(!stage.getNumChildren()createjs.Sound.play('chime');判断是否全部子蝴蝶都已经移动成功，成功则播放'chime'的声音效果。

程序运行效果如图 4-2 所示。

图 4-2 跳舞蝴蝶效果

4.3 颜色拼图游戏

1. 构建基本窗口，如代码清单 4-5 所示。

代码清单 4-5

```
<!DOCTYPE html>
<html>
<head>
<title></title>
<script src="../lib/easeljs-0.7.1.min.js"></script>
<script src="../lib/tweenjs-0.5.1.min.js"></script>
<script src="js/shapes.js"></script>
</head>
<body onload="init()">
<canvas id="canvas" width="600" height="400" style="border: black solid 1px"></canvas>
</body>
</html>
```

2. 颜色拼图效果，如代码清单 4-6 所示。

代码清单 4-6

```
var stage;
var shapes = [];
var slots = [];
var score = 0;
function init() {
    stage = new createjs.Stage("canvas");
    buildShapes();
    setShapes();
    startGame();
}
function buildShapes() {
    var colors = ['blue', 'red', 'green', 'yellow'];
    var i, shape, slot;
    for (i = 0; i < 4; i++) {
        //slots
        slot = new createjs.Shape();
        slot.graphics.beginStroke(colors[i]);
        slot.graphics.beginFill(createjs.Graphics.getRGB(255, 255, 255, 1));
        slot.graphics.drawRect(0, 0, 100, 100);
        slot.regX = slot.regY = 50;
        slot.key = i;
        slot.y = 80;
        slot.x = (i * 130) + 100;
        stage.addChild(slot);
        slots.push(slot);
        //shapes
        shape = new createjs.Shape();
```

```javascript
                shape.graphics.beginFill(colors[i]);
                shape.graphics.drawRect(0, 0, 100, 100);
                shape.regX = shape.regY = 50;
                shape.key = i;
                shapes.push(shape);
            }
        }
        function setShapes() {
            var i, r, shape;
            var l = shapes.length;
            for (i = 0; i < l; i++) {
                r = Math.floor(Math.random() * shapes.length);
                shape = shapes[r];
                shape.homeY = 320;
                shape.homeX = (i * 130) + 100;
                shape.y = shape.homeY;
                shape.x = shape.homeX;
                shape.addEventListener("mousedown", startDrag);
                stage.addChild(shape);
                shapes.splice(r, 1);
            }
        }
        function startGame() {
            createjs.Ticker.setFPS(60);
            createjs.Ticker.addEventListener("tick", function (e) {
                stage.update();
            });
        }
        function startDrag(e) {
            var shape = e.target;
            var slot = slots[shape.key];
            stage.setChildIndex(shape, stage.getNumChildren() - 1);
            stage.addEventListener('stagemousemove', function (e) {
                shape.x = e.stageX;
                shape.y = e.stageY;
            });
            stage.addEventListener('stagemouseup', function (e) {
                stage.removeAllEventListeners();
                var pt = slot.globalToLocal(stage.mouseX, stage.mouseY);
                if (slot.hitTest(pt.x, pt.y)) {
                    shape.removeEventListener("mousedown", startDrag);
                    score++;
                    createjs.Tween.get(shape).to({x:slot.x, y:slot.y}, 200, createjs.Ease.quadOut).call(checkGame);
                    console.log('h');
                }
                else {
                    createjs.Tween.get(shape).to({x:shape.homeX, y:shape.homeY}, 200, createjs.Ease.quadOut);
```

```
            }
        });
    }
    function checkGame(){
        if(score == 4){
            alert('You Win!');
        }
    }
```

❖ 代码解析

init() 函数作为函数的入口，首先创建舞台元素，接着调用 buildShapes()、setShapes()和 startGame()函数；

buildShapes()函数用 for (i = 0; i < 4; i++)循环设置四个颜色 var colors = ['blue', 'red', 'green', 'yellow']的 slot.边框矩形和 shape 实心矩形；

setShapes()函数用 for (i = 0; i < l; i++)循环 r = Math.floor(Math.random() * shapes.length)随机将其放入对应的位置，shape.addEventListener("mousedown", startDrag); startDrag 函数来响应 mousedown 事件，选定选择的实心矩形；shapes.splice(r, 1);shapes 数组减 1，直至 l 为 shapes.length 结束该函数；

startDrag(e) 函数选定选择的实心矩形，响应 stage.addEventListener('stagemousemove', function (e) { shape.x = e.stageX; shape.y = e.stageY; }); 'stagemousemove'方法将该实心矩形拖动，再响应 stage.addEventListener('stagemouseup', function (e) 的'stagemouseup'事件先 stage.removeAllEventListeners();去除所有事件，然后将该实心矩形的 var pt = slot.globalToLocal(stage.mouseX, stage.mouseY);与鼠标对应的空心矩形属性相比较，如果 slot.hitTest(pt.x, pt.y) 颜色相同则 shape.removeEventListener("mousedown",startDrag) 释放 mousedown 事件并继续运行； score++加 1；createjs.Tween.get(shape).to({x:slot.x, y:slot.y}, 200, createjs.Ease.quadOut).call(checkGame) 将该实心矩形移动空心矩形的位置重合；如果颜色不等，则 createjs.Tween.get(shape).to({x:shape.homeX, y:shape.homeY}, 200, createjs.Ease.quadOut);将空心矩形移回原来的位置；

函数 checkGame()确认当分数达到 4 时，弹出窗口显示"你成功了!!"

移动之前的效果如图 4-3 所示。

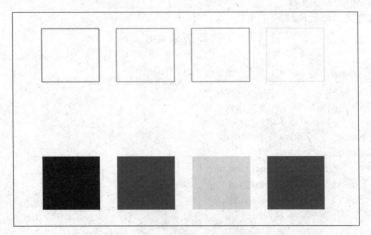

图 4-3 颜色拼图游戏初始加载效果

游戏显示成功的效果如图 4-4 所示。

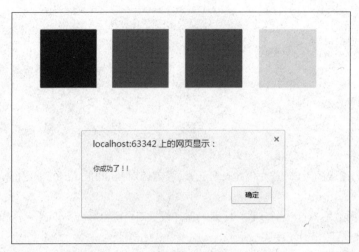

图 4-4 拼图成功效果

4.4 图像处理效果

1. 构建基本窗口，如代码清单 4-7 所示。

代码清单 4-7

```html
<!DOCTYPE html>
<html>
<head>
<title></title>
<script src="../lib/easeljs-0.7.1.min.js"></script>
<script src="../lib/preloadjs-0.4.1.min.js"></script>
<script src="../lib/filters/BoxBlurFilter.js"></script>
</head>
<body onload="preload()">
<canvas id="canvas" width="500" height="240" style="border: black solid 1px"></canvas>
</body>
```

2. 图形处理效果，如代码清单 4-8 所示。

代码清单 4-8

```javascript
<script>
    var stage, queue;
    // onload
    function preload() {
        queue = new createjs.LoadQueue();
        queue.addEventListener("complete", init);
        queue.loadManifest([
            {id:"frank", src:"img/frank.png"},
            {id:"v1", src:"img/villager1.png"},
            {id:"v2", src:"img/villager2.png"}
        ]);
    }
    function init() {
        stage = new createjs.Stage(document.getElementById('canvas'));
```

```
                sample2();
            }
        function sample2() {
            img = new Image();
            img.addEventListener('load', showFrank);
            img.src = 'img/frank.png';
        }
        function showFrank() {
            var frank = new createjs.Bitmap(img);
            stage.addChild(frank);
            stage.update();
        }
        function handleTick(e) {
            stage.update();
        }
</script>
</html>
```

❖ 代码解析

需要导入三个函数包：easeljs-0.7.1.min.js、preloadjs-0.4.1.min.js、filters/BoxBlurFilter.js；

preload()函数作为程序的主入口，主要利用 queue = new createjs.LoadQueue()来创建队列，然后 queue.addEventListener("complete", init)调用 init 函数来响应 complete 事件，接下来 queue.loadManifest 加载队列的资源 frank、v1、v2；

init()函数创建舞台并且调用不同的函数来实现不同效果，首先调用 sample2()函数；

函数 sample2() 实现通过 showFrank()来实现加载'img/frank.png'图片到舞台上，程序运行效果如图 4-5 所示。

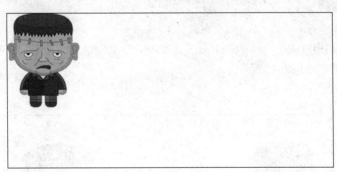

图 4-5　图形处理效果 1

修改 init()函数调用 sample()函数。

代码清单 4-9

```
function sample() {
        var frank = new createjs.Bitmap('img/frank.png');
        stage.addChild(frank);
        frank.x = 100;
        frank.y = 40;
        frank.alpha = .5;
        stage.update();
    }
```

函数 sample() 通过 var frank =createjs.Bitmap('img/frank.png');stage.addChild(frank)来实现加载

'img/frank.png'图片到舞台上，接着调整图片的高宽与透明效果。

程序运行效果如图 4-6 所示。

图 4-6　图形处理效果 2

修改 init()函数调用 drawCharacters()函数。

代码清单 4-10

```
function drawCharacters() {
    var frank = new createjs.Bitmap(queue.getResult('frank'));
    var villager1 = new createjs.Bitmap(queue.getResult('v1'));
    var villager2 = new createjs.Bitmap(queue.getResult('v2'));
    frank.y = villager1.y = villager2.y = 40;
    frank.x = 20;
    villager1.x = 190;
    villager2.x = 360;
    stage.addChild(frank, villager1, villager2);
    stage.update();
}
```

drawCharacters()函数通过 createjs.Bitmap()在舞台上不同的位置加载三个图像 frank、villager1、villager2，然后调整三个图像的位置。

程序运行效果如图 4-7 所示。

图 4-7　图形处理效果 3

修改 drawCharacters()函数加上如下代码设置阴影的颜色、角度与粗细效果。

代码清单 4-11

```
villager1.shadow = new createjs.Shadow('#000', 5, 5, 8);
```

程序运行效果如图 4-8 所示。

图 4-8 图形处理效果 4

修改 drawCharacters() 函数。

代码清单 4-12

```
var w = frank.image.width;
    var h = frank.image.height;
    villager1.shadow = new createjs.Shadow('#000', 5, 5, 8);
    var blur = new createjs.BoxBlurFilter(5, 5, 1);
    frank.filters = [blur];
    frank.cache(0, 0, w, h);
    stage.update();
}
```

new createjs.BoxBlurFilter(5, 5, 1)设置高斯模糊滤镜过滤器；frank.filters = [blur]设置 frank 的过滤效果；frank.cache(0, 0, w, h)以缓存的方式保证以后的调用能持续生效。

程序运行效果如图 4-9 所示。

图 4-9 图形处理效果 5

修改 drawCharacters() 函数添加如下代码。

代码清单 4-13

```
var color = new createjs.ColorFilter(1, 0, 0, 1, 0, 0, 0, 0);
villager2.filters = [color, blur];
villager2.cache(0, 0, w + 10, h + 10);
    stage.update();
}
```

new createjs.ColorFilter(1, 0, 0, 1, 0, 0, 0, 0);设置颜色过滤器；villager2.filters = [color, blur];设置 villager2 的过滤效果。

程序运行效果如图 4-10 所示。

图 4-10　图形处理效果 6

4.5　处理跑跳效果

1. 构建基本窗口，如代码清单 4-14 所示。

代码清单 4-14

```html
<!DOCTYPE html>
<html>
<head>
<title></title>
<script src="../lib/easeljs-0.7.1.min.js"></script>
<script src="../lib/preloadjs-0.4.1.min.js"></script>
<script src="Runner.js"></script>
<script src="SimpleButton.js"></script>
<script src="Preloader.js"></script>
<script src="runningMan.js"></script>
</head>
<body onload="preload()">
<canvas id="canvas" width="800" height="600" style="border: black solid 1px;background-color: #add8e6"></canvas>
</body>
</html>
```

2. 定义 RunningMan.js 来实现跑跳页面初始化设计。

代码清单 4-15

```javascript
var stage, queue, preloader, spritesheet, runner;
function preload() {
    queue = new createjs.LoadQueue();
    queue.loadManifest([
        {id:"runner", src:"img/runningMan.png"}
    ],false);
    init();
}
function init(){
    stage = new createjs.Stage(document.getElementById('canvas'));
    createjs.Ticker.on('tick', stage);
```

```
        stage.enableMouseOver();
        preloader = new ui.Preloader('#FFF','#000');
        preloader.x = (stage.canvas.width / 2) - (preloader.width / 2);
        preloader.y = (stage.canvas.height / 2) - (preloader.height / 2);
        stage.addChild(preloader);
        queue.addEventListener("complete", initGame);
        queue.addEventListener('progress', onFileProgress);
        queue.load();
    }
    function onFileProgress(e) {
        preloader.update(e.progress);
    }
    function initGame() {
        stage.removeChild(preloader);
        preloader = null;
        spritesheet = new createjs.SpriteSheet({
            "images":[queue.getResult("runner")],
            "frames":{"regX":0, "height":292, "count":64, "regY":0, "width":165},
            "animations":{"idle":[60], "run":[0, 25], "jump":[31, 60, 'idle']}
        });
        buildRunner();
        buildButtons();
    }
    function buildRunner() {
        runner = new sprites.Runner(spritesheet);
        runner.y = 100;
        stage.addChild(runner);
    }
    function buildButtons() {
        var jumpBtn = new ui.SimpleButton("JUMP");
        var runBtn = new ui.SimpleButton("RUN");
        var idleBtn = new ui.SimpleButton("IDLE");
        jumpBtn.on('click', function (e) {
            runner.jump();
        });
        runBtn.on('click', function (e) {
            runner.run();
        });
        runBtn.x = jumpBtn.width + 10;
        idleBtn.on('click', function (e) {
            runner.stand();
        });
        idleBtn.x = runBtn.x + runBtn.width + 10;
        stage.addChild(jumpBtn, runBtn, idleBtn);
    }
}
```

3. 用 CreateJS 来实现跑跳效果，如代码清单 4-16 所示。

代码清单 4-16

```
(function () {
    window.sprites = window.sprites || {};
    var Runner = function (spritesheet) {
```

```
            this.initialize(spritesheet);
        }
        var p = Runner.prototype = new createjs.Sprite();
        p.Sprite_initialize = p.initialize;
        p.speed = 0;
        p.initialize = function (spritesheet) {
            this.Sprite_initialize(spritesheet, 'idle');
            this.on('tick', function (e) {
                this.x += this.speed;
                if (this.x > stage.canvas.width) {
                    this.x = -this.getBounds().width;
                }
            })
        }
        p.run = function () {
            if (this.currentAnimation === 'idle') {
                this.gotoAndPlay('run');
                this.speed = 10;
            }
        }
        p.jump = function () {
            if (this.currentAnimation != 'jump') {
                this.gotoAndPlay('jump');
                this.on('animationend', function (e) {
                    if (this.speed > 0) {
                        this.gotoAndPlay('run');
                    }
                })
            }
        }
        p.stand = function () {
            if (this.currentAnimation === 'run') {
                this.gotoAndStop('idle');
                this.speed = 0;
            }
        }
        window.sprites.Runner = Runner;
}());
```

◆ 代码解析

需要导入以下函数包：easeljs-0.7.1.min.js、preloadjs-0.4.1.min.js、Runner.js、SimpleButton.js、Preloader.js、runningMan.js 来实现跑跳效果；

首先需要了解一下 RunningMan 图片运动的效果定义方式，这个主要是放在 runner.js 的函数块中：通过 window.sprites = window.sprites || {}; var Runner = function (spritesheet) { this.initialize (spritesheet);}来扩展系统的 sprites 函数效果；var p = Runner.prototype = new createjs.Sprite();创建 p 为 runner 的对象实例，然后定义这个对象实例的初始化选项，设置其运行的默认速度等；接着分别定义 p.initialize、p.run、p.jump、p.stand 来实现初始化、跑、跳与听的动画切片与速度效果；

Preload()函数作为程序的主入口，主要利用 queue = new createjs.LoadQueue()来创建队列，接着用 queue.loadManifest 把 runner 整张大图的资源加载入队列，然后 init 调用 init 函数；

整张大图如图 4-11 所示。

图 4-11 跑跳帧资源图

init()函数创建舞台,设置舞台的背景颜色、大小等,再依次相应两个事件:addEventListener("complete", initGame);舞台准备好之后响应 initGame 事件;.addEventListener('progress', onFileProgress)程序运行时响应 onFileProgress 事件 preloader.update(e.progress)来进度条实时更新进度;

程序运行效果如图 4-12 所示。

图 4-12 进度条

initGame() 函数清除默认的背景效果,spritesheet = new createjs.SpriteSheet 创建 spriteSheet 的图片来源、剪切规格和 idle、run 和 jump 的效果图,然后调用 buildRunner()和 buildButtons()函数;

buildRunner()调用 runner.js 来创建 runner 对象实例,并设置其 y 轴为 100,将 runner 对象添加到舞台上;

buildButtons()函数调用 SimpleButton.js 的 new ui.SimpleButton 函数库创建简单的 JUMP、RUN 和 IDLE 按钮。

程序实现效果如图 4-13 所示。

当 jumpBtn 被点击时,相应的 runner.jump()效果如图 4-14 所示。

当 runBtn 被点击时,相应的 runner.run()效果如图 4-15 所示。

当 jump idleBtn Btn 被点击时,相应的 runner. stand ()效果如图 4-16 所示。

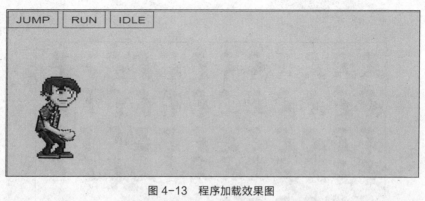

图 4-13 程序加载效果图

图 4-14 跳跃效果图

图 4-15 跑效果图

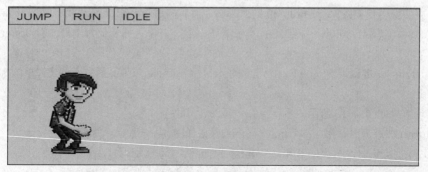

图 4-16 静止效果图

4.6 实现炫酷效果

本课主要讲解应用 Canvas 与 CreateJS 制作出炫酷效果。

极客学院在线视频学习网址：

http://www.jikexueyuan.com/course/229_2.html

手机扫描二维码

使用 Canvas 制作炫酷的效果

1. 构建基本窗口，如代码清单 4-17 所示。

代码清单 4-17

```
<!DOCTYPE html>
<html>
<head lang="en">
<meta charset="UTF-8">
<title></title>
<script src="easeljs-0.7.1.min.js"></script>
</head>
<body>
<canvas id="canvas" width="1000px" height="500px"></canvas>
<script src="app.js"></script>
</body>
</html>
```

2. 构建程序主要功能，如代码清单 4-18 所示。

代码清单 4-18

```
var canvas;
var stage;
var img = new Image();
var sprite;
window.onload = function(){
    canvas = document.getElementById("canvas");
    stage = new createjs.Stage(canvas);
    stage.addEventListener("stagemousedown",clickCanvas);
    stage.addEventListener("stagemousemove",moveCanvas);
    var data={
        images:["2.png"],
        frames:{width:20,height:20,regX:10,regY:10}
    }
    sprite   = new createjs.Sprite(new createjs.SpriteSheet(data));
    createjs.Ticker.setFPS(20);
    createjs.Ticker.addEventListener("tick",tick);
}
```

```
function tick(e){
    var t = stage.getNumChildren();
    for(var i = t-1;i>0;i--){
        var s = stage.getChildAt(i);
        s.vY +=2;
        s.vX +=1;
        s.x += s.vX;
        s.y += s.vY;
        s.scaleX = s.scaleY =s.scaleX+ s.vS;
        s.alpha += s.vA;
        if(s.alpha <= 0 || s.y >canvas.height){
            stage.removeChildAt(i);
        }
    }
    stage.update(e);
}
function clickCanvas(e){
    addS(Math.random()*200 + 100,stage.mouseX,stage.mouseY,2);
}
function moveCanvas(e){
    addS(Math.random()*2 + 1,stage.mouseX,stage.mouseY,1);
}
function addS(count,x,y,speed){
    for(var i = 0;i<count;i++){
        var s = sprite.clone();
        s.x = x;
        s.y = y;
        s.alpha = Math.random()*0.5 + 0.5;
        s.scaleX = s.scaleY = Math.random() +0.3;
        var a = Math.PI * 2 *Math.random();
        var v = (Math.random() - 0.5) *30 *speed;
        s.vX = Math.cos(a) *v;
        s.vY = Math.sin(a) *v;
        s.vS = (Math.random() - 0.5) *0.2; // scale
        s.vA = -Math.random() *0.05 -0.01; // alpha
        stage.addChild(s);
    }
}
```

✧ 代码解析

需要导入两个函数包：easeljs-0.7.1.min.js、app.js 来实现炫酷效果。

加载 app.js 函数，window.onload = function()作为主要函数的入口，首先创建舞台基本属性，再用 addEventListener("stagemousedown",clickCanvas) 在鼠标按下时候响应 clickCanvas 函数，addEventListener("stagemousemove",moveCanvas)在鼠标移动时响应 moveCanvas 函数；然后用定义 data、new createjs.Sprite(new createjs.SpriteSheet(data))方法将图片 2.png 加载入页面，设置页面的刷新频率为 20 帧/秒，最后再去响应 tick 事件函数。

初始化之后程序运行结果如图 4-17 所示。

addS(count,x,y,speed)有四个参数，其中 count 指明炫酷数量，x、y 为 x/y 轴炫酷范围、speed 来制定炫酷的速度；通过一个 for(var i = 0;i<count;i++)循环，s = sprite.clone()获得实例对象，设置其出现的位置（x、y）、用随机数设置透明度 alpha 的变化、用随机数设置 scaleX 缩放效果、设置活动的

曲线变化和滑动的速度（vX、vY）、用随机数来设置滑动的图形 vS 的 scaleX 缩放、用随机数来设置滑动的图形 vA 的 alpha 透明度的变化；

图 4-17　程序背景图

clickCanvas()函数调用 addS(Math.random()*200 + 100,stage.mouseX,stage.mouseY,2)函数来规定炫酷效果的属性；

moveCanvas()函数调用 addS(Math.random()*2 + 1,stage.mouseX,stage.mouseY,1); 函数来规定炫酷效果的属性，对比 clickCanvas()，可见点击的炫酷效果更加绚丽；

tick(e)函数通过 stage.getNumChildren()得到设置的所有子对象，通过 for(var i = t-1;i>0;i--)循环一个个地取出来，得到当前对象 var s = stage.getChildAt(i)，然后设置这个对象的移动坐标变化、缩放效果和透明度变化情况，当(s.alpha <= 0 || s.y >canvas.height)透明度≤0 或者 y 轴在舞台底边的时候，不再显示，清除该元素；更新舞台。

鼠标点击后的程序运行结果如图 4-18 所示。

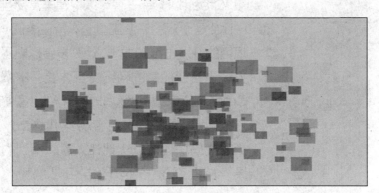

图 4-18　鼠标点击后的炫酷效果

鼠标移动后的程序运行结果如图 4-19 所示。

图 4-19　鼠标移动后的炫酷效果

第5章

HTML5小型游戏

■ CreateJS 是一套可以构建丰富交互体验的 HTML5 游戏的开源工具包，旨在降低 HTML5 项目的开发难度和成本，让开发者以熟悉的方式打造更具现代感的网络交互体验。本章主要用 CreateJS 来编写两款非常流行的小型游戏：围住神经猫和看你有多色游戏，可以看出，只要很少的代码块就可以实现让人意想不到的游戏效果，心动了吧，一起开始学习吧。

5.1 围住神经猫游戏

围住神经猫的游戏是一款在网络上非常火爆的游戏，需要用最少的步数将神经猫围起来，恶搞又益智。

核心内容：
1. 围住神经猫游戏玩法。
2. 用 CreateJS 进行 HTML5 游戏开发。
3. 围住神经猫游戏的具体开发流程。

开发环境：IDEA

5.1.1 介绍围住神经猫游戏的玩法

极客学院在线视频学习网址：
http://www.jikexueyuan.com/course/158_1.html
手机扫描二维码

围住神经猫游戏的玩法介绍

图 5-1 所示是其初始界面。

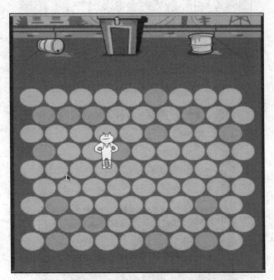

图 5-1 围住神经猫游戏界面

神经猫游戏界面上只有两种颜色，猫只能走灰色的点，而用鼠标点击灰色点后就会变成橙色的点，只要点到橙色部分点围住小猫就算成功，如图 5-2 所示。

如图 5-3 所示，每次围住小猫后都会弹出一个系统全国排名比分，看看用了几步可将小猫围住。

如图 5-4 所示，如果没有围住，则跳出图 5-4 所示对话框。

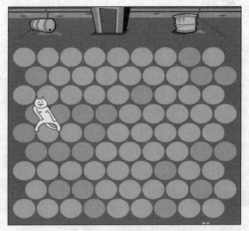

图 5-2　神经猫被围住

图 5-3　围住神经猫之后的分享界面

图 5-4　没有围住神经猫游戏对话框

本节 HTML5 围住神经猫游戏网页版是一款基于 HTML5 canvas、egret_loader.js 和 jquery.min.js、CreateJS 等技术制作的游戏，该代码都是开源的，对学习网页游戏制作的学员有非常好的借鉴作用。

5.1.2 使用 CreateJS 围住神经猫

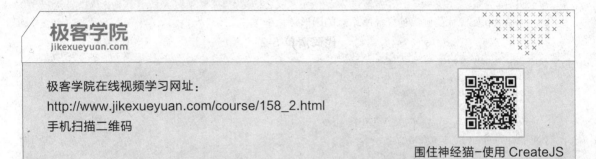

CreateJS 在 www.createjs.com 提供了 EaselJS、TwenJS、SoundJS 和 PreloadJS 的下载，前面章节已经做过详细的说明，神经猫开发只用到 EaselJS，点击主页的 EaselJS 链接，EaselJS 库文件下载界面如图 5-5 所示。

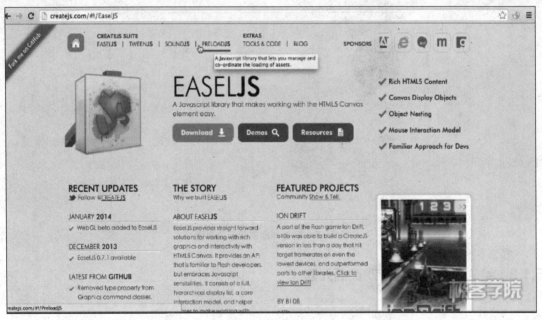

图 5-5　Easel 下载地址

1. 下载完之后在源代码中加载。

代码清单 5-1

```
<!DOCTYPE html>
<html>
<head lang="en">
<meta charset="UTF-8">
<title></title>
<script src="easeljs-0.7.1.min.js"></script>
```

```
</head>
<body>
<canvas width="800px" height="800px" id="gameView"></canvas>
<script src="app1.js"></script>
</body>
</html>
```

2. 编写 app.js 源代码画出神经猫所站立的圆形测试脚本。

代码清单 5-2

```
var stage = new createjs.Stage("gameView");
var gameView = new createjs.Container();
stage.addChild(gameView);

var s = new createjs.Shape();
s.graphics.beginFill("#ff0000");
s.graphics.drawCircle(50,50,25);
s.graphics.endFill();
gameView.addChild(s);
createjs.Ticker.setFPS(30);
createjs.Ticker.addEventListener("tick",stage);
```

3. 程序运行的效果如图 5-6 所示。

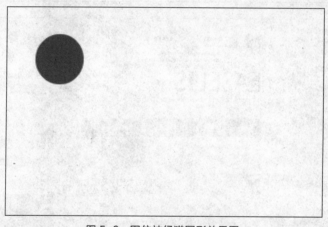

图 5-6 围住神经猫圆形效果图

5.1.3 绘制围住神经猫游戏页面元素

极客学院在线视频学习网址：

http://www.jikexueyuan.com/course/158_3.html

手机扫描二维码

围住神经猫游戏页面元素的绘制

本节主要学习 HTML5 游戏围住神经猫 UI 的绘制。这里没有神经猫的图片，所以不引入这张图片，而用另外一张带颜色的圆形图片代表神经猫。

1. 增加主要的<script src="Circle.js"></script>代码，修改<script src="app.js"></script>来绘制神经猫游戏的圆形背景与神经猫效果。

2. app.js 中添加如下代码。

代码清单 5-3

```
var stage = new createjs.Stage("gameView");
createjs.Ticker.setFPS(30);
createjs.Ticker.addEventListener("tick",stage);
var gameView = new createjs.Container();
stage.addChild(gameView);
```

添加舞台 stage，setFPS(30)刷新频率为 30 帧/秒，createjs.Container();创建一个 container 对象来设置舞台元素。

3. 接着创建 Circle.js 文件来绘制圆形形状。

代码清单 5-4

```
function Circle(){
    createjs.Shape.call(this);
    this.setCircleType = function(type){
        this._circleType = type;
        switch (type){
            case 1:
                this.setColor("#cccccc");
                break;
            case 2:
                this.setColor("#ff6600");
                break;
            case 3:
                this.setColor("#0000ff");
                break;
        }
    }
    this.setColor = function (colorString){
        this.graphics.beginFill(colorString);
        this.graphics.drawCircle(0,0,25);
        this.graphics.endFill();
    }
    this.getCircleType = function(){
        return this._circleType;
    }
    this.setCircleType(1);
}
Circle.prototype = new createjs.Shape();
```

function Circle()用面向对象的方法来定义 Circle 类的一个实体，createjs.Shape.call(this)回调一下这个方法，setCircleType 方法设置圆的三种颜色：1 为 ("#cccccc")、2 为("#ff6600")、3 为("#0000ff")；setColor 方法绘制指定颜色的圆形；getCircleType 得到圆形的颜色； setCircleType(1)默认的圆形颜色为("#cccccc")。

4. 继续添加 app.js 代码。

代码清单 5-5

```
gameView.x = 30;
gameView.y = 30;
var circleArr = [[],[],[],[],[],[],[],[],[]];
function addCircles(){
    for(var indexY=0;indexY<9;indexY++){
        for(var indexX = 0;indexX<9;indexX++){
            var c = new Circle();
            gameView.addChild(c);
            circleArr[indexX][indexY] = c;
            c.indexX = indexX;
            c.indexY = indexY;
            c.x = indexX*55;
            c.y = indexY*55;
        }
    }
}
addCircles();
```

用 circleArr 数组来保存圆形背景效果，通过 addCircles()函数用 for(var indexY=0;indexY<9; indexY++){ for(var indexX = 0;indexX<9;indexX++)}循环来设置 9×9 个圆形，效果如图 5-7 所示。

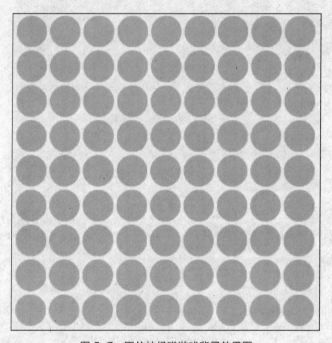

图 5-7　围住神经猫游戏背景效果图

5. 继续在 5×5 的地方绘制猫，在 for 循环中添加如下代码。

代码清单 5-6

```
if(indexX==4&&indexY==4){
    c.setCircleType(3);
    currentCat = c;
}}
```

效果如图 5-8 所示。

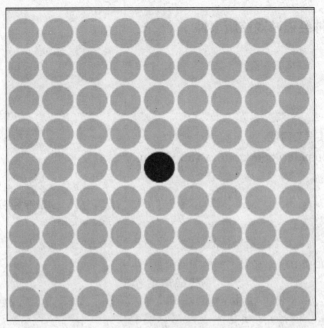

图 5-8　围住神经猫猫效果图

6. 需要修改 9×9 圆形中，让其单数行不变，双数行缩进 25，修改 c.x 的代码如下，效果如图 5-9 所示。

代码清单 5-7

```
c.x = indexY%2?indexX*55+25:indexX*55;
```

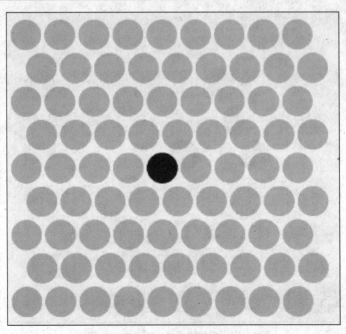

图 5-9　围住神经猫圆形背景缩进效果图

这样整个神经猫的背景效果和猫的显示效果都已经完成了。

5.1.4 添加围住神经猫游戏监听事件

极客学院在线视频学习网址：
http://www.jikexueyuan.com/course/158_4.html
手机扫描二维码

围住神经猫游戏监听事件的添加

本节给围住神经猫游戏添加监听事件的处理。

1. 在 addCircle()函数的 for 循环内部添加监听事件。

代码清单 5-8

```
c.addEventListener("click",function(event){
if(event.target.getCircleType() != 3){
    event.target.setCircleType(2);
}
```

该事件监听鼠标点击动作，调用 SetCircleType()函数将该圆的背景改为("#ff6600")；添加了 if(event.target.getCircleType() != 3) 防止当点击猫时，猫的颜色也发生改变。

效果显示如图 5-10 所示。

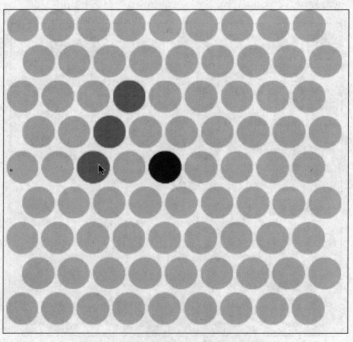

图 5-10　围住神经猫鼠标点击改变颜色

2. 将该事件方法提炼成一个函数 circleClicked(event)。

代码清单 5-9

```
c.addEventListener("click",circleClicked);
function circleClicked(event){
    if(event.target.getCircleType() != Circle.TYPE_CAT ){
        event.target.setCircleType(Circle.TYPE_SELECTED);
    }
else{return;}
}
```

5.1.5 使用简单的逻辑实现围住神经猫游戏效果

极客学院在线视频学习网址：
http://www.jikexueyuan.com/course/158_5.html
手机扫描二维码

简单的逻辑实现围住神经猫游戏效果

本节讲解围住神经猫简单逻辑游戏实现，点击圆形时，想让猫移动起来。

1. 添加变量 var currentCat;保存这只猫。
2. 在 addCircle()函数体中将猫赋值给当前点击的对象。

代码清单 5-10

```
if(indexX==4&&indexY==4){
    c.setCircleType(3);
    currentCat = c;
}
```

3. 对当前 circleClicked(event)函数体中添加猫跳跃判断。

代码清单 5-11

```
if(currentCat.indexX == 0 || currentCat.indexX==8 ||currentCat.indexY == 0 || currentCat.indexY==8){
    alert("游戏结束");
    return;
}
```

当 if(currentCat.indexX == 0 || currentCat.indexX==8 ||currentCat.indexY == 0 ||currentCat.indexY==8)猫跑到边缘时，弹出图 5-11 所示对话框，游戏结束。

4. 用 switch 语句判定猫跳跃效果。

代码清单 5-12

```
var leftCircle = circleArr [ currentCat.indexX -1][currentCat.indexY]
var rightCircle = circleArr [ currentCat.indexX +1][currentCat.indexY]
var lefttopCircle = circleArr [ currentCat.indexX -1][currentCat.indexY-1]
var righttopCircle = circleArr [ currentCat.indexX][currentCat.indexY-1]
var leftbottomCircle = circleArr [ currentCat.indexX -1][currentCat.indexY+1]
var rightbottomCircle = circleArr [ currentCat.indexX][currentCat.indexY+1]
if(leftCircle.getCircleType() == 1){
    leftCircle.setCircletype(3);
```

```
        currentCat.setCircletype(1);
        currentCat = leftCircle
    }else if(rightCircle.getCircleType() == 1){
        rightCircle.setCircletype(3);
        currentCat.setCircletype(1);
        currentCat = rightCircle
    }else if(lefttopCircle.getCircleType() == 1){
        lefttopCircle.setCircletype(3);
        currentCat.setCircletype(1);
        currentCat = lefttopCircle
    } else if(righttopCircle.getCircleType() == 1){
        righttopCircle.setCircletype(3);
        currentCat.setCircletype(1);
        currentCat = righttopCircle
    }else if (leftbottomCircle.getCircleType() == 1){
        leftbottomCircle.setCircletype(3);
        currentCat.setCircletype(1);
        currentCat = leftbottomCircle
    } else if(rightbottomCircle.getCircleType() == 1){
        rightbottomCircle.setCircletype(3);
        currentCat.setCircletype(1);
        currentCat = rightbottomCircle;
    else {
        alert("游戏结束");
    }
```

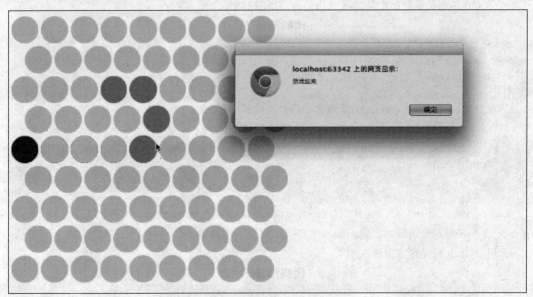

图 5-11　弹出游戏结束对话框

用 switch 语句判定猫移动的方向总共有六个方向：leftCircle、rightCircle、lefttopCircle、righttopCircle、leftbottomCircle 和 leftbottomCircle，当此方向的圆形未使用时，设置该方向为猫跳跃的方向，而原来猫的位置改为未使用状态；当所有的方向都走不通时，弹出对话框，游戏结束。

代码实现效果如图 5-12 所示。

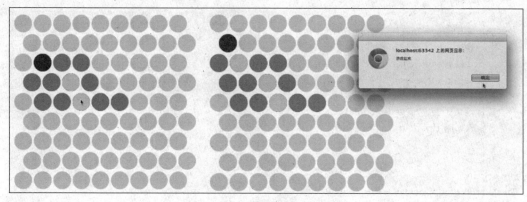

图 5-12　围住神经猫游戏效果图

到此，已经基本完成了猫跳跃功能的实现，下节对代码进行整合和优化。

5.1.6　实现围住神经猫游戏完整效果

极客学院在线视频学习网址：
http://www.jikexueyuan.com/course/158_6.html
手机扫描二维码

围住神经猫游戏完整效果实现

本节完整地开发围住神经猫游戏，实现逻辑内容与代码优化。

1．优化 Circle.js 代码，将 1、2、3 改为公共的参数 Circle.TYPE_UNSELECTE、Circle.TYPE_SELECTED、Circle.TYPE_CAT。

代码清单 5-13

```
function Circle(){
    createjs.Shape.call(this);
    this.setCircleType = function(type){
        this._circleType = type;
        switch (type){
            case Circle.TYPE_UNSELECTED:
                this.setColor("#cccccc");
                break;
            case Circle.TYPE_SELECTED:
                this.setColor("#ff6600");
                break;
            case Circle.TYPE_CAT:
                this.setColor("#0000ff");
                break;
        }
    }
```

```
        this.setColor = function (colorString){
            this.graphics.beginFill(colorString);
            this.graphics.drawCircle(0,0,25);
            this.graphics.endFill();
        }
        this.getCircleType = function(){
            return this._circleType;
        }
        this.setCircleType(1);
}
Circle.prototype = new createjs.Shape();
Circle.TYPE_UNSELECTED = 1;
Circle.TYPE_SELECTED = 2;
Circle.TYPE_CAT = 3;
```

2. 优化后的 app.js 代码如下所示。

代码清单 5-14

```
var stage = new createjs.Stage("gameView");
createjs.Ticker.setFPS(30);
createjs.Ticker.addEventListener("tick",stage);
var gameView = new createjs.Container();
gameView.x = 30;
gameView.y = 30;
stage.addChild(gameView);
var circleArr = [[],[],[],[],[],[],[],[],[]];
var currentCat;
var MOVE_NONE=-1,MOVE_LEFT = 0,MOVE_UP_LEFT=1,MOVE_UP_RIGHT= 2,MOVE_RIGHT=3,MOVE_DOWN_RIGHT= 4,MOVE_DOWN_LEFT=5;
function getMoveDir(cat){
    var distanceMap=[];
    //left
    var can = true;
    for(var x = cat.indexX;x>=0;x--){
        if(circleArr[x][cat.indexY].getCircleType() == Circle.TYPE_SELECTED){
            can = false;
            distanceMap[MOVE_LEFT] = cat.indexX-x;
            break;
        }
    }
    if(can){
        return MOVE_LEFT;
    }
    //left up
    can = true;
    var x = cat.indexX, y = cat.indexY;
    while(true){
        if(circleArr[x][y].getCircleType() == Circle.TYPE_SELECTED){
            can = false;
            distanceMap[MOVE_UP_LEFT] = cat.indexY-y;
            break;
        }
```

```
            if(y%2 == 0){
                x--;
            }
            y--
            if(y<0||x<0){
                break;
            }  }
    if(can){
        return MOVE_UP_LEFT;
    }
    //right up
    can = true;
    x = cat.indexX,y = cat.indexY
    while(true){
        if(circleArr[x][y].getCircleType() == Circle.TYPE_SELECTED){
            can = false;
            distanceMap[MOVE_UP_RIGHT] = cat.indexY - y;
            break;
        }
        if(y%2 ==1){
            x++
        }
        y--;
        if(y<0 || x>8){
            break
        } }
    if(can){
        return MOVE_UP_RIGHT;
    }
    //right
    can = true;
    for(var x = cat.indexX;x<9;x++){
        if(circleArr[x][cat.indexY].getCircleType() == Circle.TYPE_SELECTED){
            can = false;
            distanceMap[MOVE_RIGHT] = x - cat.indexX;
            break;
        }   }
    if(can){
        return MOVE_RIGHT;
    }
    //right down
    can = true;
    x = cat.indexX,y = cat.indexY;
    while(true){
        if(circleArr[x][y].getCircleType() == Circle.TYPE_SELECTED){
            can = false;
            distanceMap[MOVE_DOWN_RIGHT] = y-cat.indexY;
            break;
        }
        if(y%2 == 1){
            x++;
```

```
            }
            y++;
            if(y>8 ||x>8){
                break;
            }   }
        if(can){
            return MOVE_DOWN_RIGHT;
        }
        //left down
        can = true;
        x = cat.indexX,y=cat.indexY;
        while(true){
            if(circleArr[x][y].getCircleType() == Circle.TYPE_SELECTED){
                can = false;
                distanceMap[MOVE_DOWN_LEFT] = y-cat.indexY;
                break;
            }
            if(y%2==0){
                x--;
            }
            y++;
            if(y>8||x<0){
                break;
            }   }
        if(can){
            return MOVE_DOWN_LEFT;
        }
        var maxDir = -1,maxValue= -1;
        for(var dir = 0;dir <distanceMap.length;dir++){
            if(distanceMap[dir]>maxValue){
                maxValue = distanceMap[dir];
                maxDir = dir;
            }   }
        if(maxValue>1){
            return maxDir;
        }else{
            return MOVE_NONE;
        }}
function circleClicked(event){
    if(event.target.getCircleType() != Circle.TYPE_CAT ){
        event.target.setCircleType(Circle.TYPE_SELECTED);
    }else
    {
        return;
    }
    if(currentCat.indexX == 0 || currentCat.indexX==8 ||currentCat.indexY == 0 || currentCat.indexY==8){
        alert("游戏结束");
        return;
    }
```

```javascript
            var dir = getMoveDir(currentCat)
            switch (dir){
                case MOVE_LEFT:
                    currentCat.setCircleType(Circle.TYPE_UNSELECTED);
                    currentCat = circleArr[currentCat.indexX-1][currentCat.indexY];
                    currentCat.setCircleType(Circle.TYPE_CAT);
                    break;
                case MOVE_UP_LEFT:
                    currentCat.setCircleType(Circle.TYPE_UNSELECTED);
                    currentCat =
 circleArr[currentCat.indexY%2?currentCat.indexX:currentCat.indexX-1][currentCat.indexY-1];
                    currentCat.setCircleType(Circle.TYPE_CAT);
                    break;
                case MOVE_UP_RIGHT:
                    currentCat.setCircleType(Circle.TYPE_UNSELECTED);
                    currentCat =
circleArr[currentCat.indexY%2?currentCat.indexX+1:currentCat.indexX][currentCat.indexY-1];
                    currentCat.setCircleType(Circle.TYPE_CAT);
                    break;
                case MOVE_RIGHT:
                    currentCat.setCircleType(Circle.TYPE_UNSELECTED);
                    currentCat = circleArr[currentCat.indexX+1][currentCat.indexY];
                    currentCat.setCircleType(Circle.TYPE_CAT);
                    break;
                case MOVE_DOWN_RIGHT:
                    currentCat.setCircleType(Circle.TYPE_UNSELECTED);
                    currentCat =
circleArr[currentCat.indexY%2?currentCat.indexX+1:currentCat.indexX][currentCat.indexY+1];
                    currentCat.setCircleType(Circle.TYPE_CAT);
                    break;
                case MOVE_DOWN_LEFT:
                    currentCat.setCircleType(Circle.TYPE_UNSELECTED);
                    currentCat =
circleArr[currentCat.indexY%2?currentCat.indexX:currentCat.indexX-1][currentCat.indexY+1];
                    currentCat.setCircleType(Circle.TYPE_CAT);
                    break;
                default :
                    alert("游戏结束");
            }}
    function addCircles(){
        for(var indexY=0;indexY<9;indexY++){
            for(var indexX = 0;indexX<9;indexX++){
                var c = new Circle();
                gameView.addChild(c);
                circleArr[indexX][indexY] = c;
                c.indexX = indexX;
                c.indexY = indexY;
                c.x = indexY%2?indexX*55+25:indexX*55;
                c.y = indexY*55;

                if(indexX==4&&indexY==4){
```

```
                c.setCircleType(3);
                currentCat = c;
            }else if(Math.random()<0.1){
                c.setCircleType(Circle.TYPE_SELECTED);
            }
            c.addEventListener("click",circleClicked);
        } }}
addCircles();
```

3. 其中设置猫移动的六个方向的参数是 MOVE_LEFT、MOVE_UP_LEFT、MOVE_UP_RIGHT、MOVE_RIGHT、MOVE_DOWN_RIGHT、MOVE_DOWN_LEFT，无处可移设置为 MOVE_NONE。

4. 修改 circleClicked(event)函数，添加 getMoveDir(cat)函数分左、左上、右上、右下、左下、右下方向判断猫可移动的方向。

5. circleClicked(event)函数获取到猫可移动的方向 var dir = getMoveDir(currentCat)之后，用 Switch 语句设置该方向为猫跳跃的方向，而原来猫的位置改为未使用状态；当所有的方向都走不通时，弹出对话框游戏结束。

6. 到此为止神经猫游戏已经基本完成，运行之后，刚载入页面时效果如图 5-13 所示。

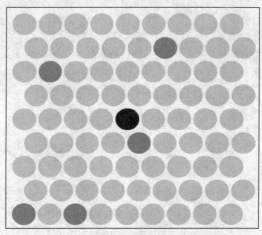

图 5-13　加载围住神经猫游戏初始随机界面

围住神经猫后效果如图 5-14 所示。

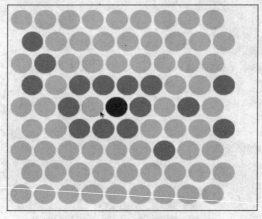

图 5-14　成功围住神经猫效果图

失败后页面效果如图 5-15 所示。

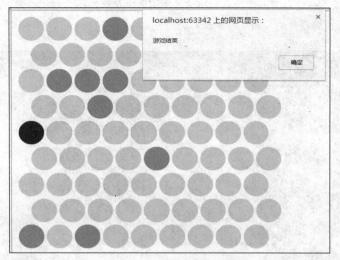

图 5-15　围住神经猫失败效果图

5.2　看你有多色游戏

由深圳市东方博雅科技有限公司独立研发，一款名为"看你有多色"的小游戏风靡朋友圈，上线一天就以单日 1013 万注册用户的傲人成绩，成为移动互联网当之无愧的 MVP。

核心内容：

1. 了解看你有多色游戏。
2. 掌握 CreateJS 绘制图形。
3. 掌握游戏的开发。

开发环境：Eclipse

5.2.1　介绍看你有多色游戏

极客学院在线视频学习网址：
http://www.jikexueyuan.com/course/167_1.html
手机扫描二维码

看你有多色-游戏介绍

"看你有多色"是一款基于 HTML5 技术、挑战人类眼球对颜色的分辨能力、好玩易上手的小游戏。游戏画面非常简单，基本由方格组合而成，弧形的边框体现出了开发商对细节方面的考究，给玩家带来一种非常舒适的视觉体验，同色系的色块用白色的线条区分开来，粗细适中，使得没有一大片同色系连在一起的错觉感受。游戏根据移动互联网的用户行为习惯进行简化,仅通过色块的颜色差距来进行"找色";随着游戏时间的深入,色块的数量不断增加,相似度也更加靠近,对玩家眼神的考验更加严峻,富有挑战性。看你有多色游戏开始页面如图 5-16 所示。

图 5-16　看你有多色游戏界面

点击图 5-16 游戏界面中的"开始游戏"进入图 5-17 的游戏主界面,四方格中有三个颜色一样,每个颜色都不一样。

图 5-17　看你有多色游戏 2×2 方格主界面

点击不一样颜色的方格,将出现图 5-18 所示的更多方格效果,且颜色也变得更加相似,以此类推,直到 64 个方格后不再增加更多方格,只改变颜色使其颜色相似度越来越高。

图 5-18 看你有多色游戏方格界面

5.2.2 使用 CreateJS 开发看你有多色

极客学院在线视频学习网址：
http://www.jikexueyuan.com/course/167_2.html
手机扫描二维码

看你有多色-CreateJS 介绍

1. 前面章节已经详细地说明过如何下载 CreateJS 包文件，下载完之后在源代码中加载。神经猫开发只用到 EaselJS。

代码清单 5-15

```
<!DOCTYPE html>
<html>
<head lang="en">
<meta charset="UTF-8">
<meta name="viewport" content="width=device-width,user-scalable=no">
<script src="easeljs-0.7.1.min.js"></script>
</head>
<body>
</div>
<div class="main">
<canvas id="gameView"></canvas>
</div>
<script src="app.js"></script>
</body>
</html>
```

2. 编写 app.js 实现绘制矩形的测试代码。

代码清单 5-16

```
var stage = new createjs.Stage("gameView");
var gameView = new createjs.Container();
stage.addChild(gameView);
var s = new createjs.Shape();
s.graphics.beginFill("#ff0000");
s.graphics.drawRect(0,0,100,100);
s.graphics.endFill();
gameView.addChild(s);
createjs.Ticker.setFPS(30);
createjs.Ticker.addEventListener("tick",stage);
```

3. 程序运行的效果如图 5-19 所示。

图 5-19　实现方格效果图

5.2.3　制作看你有多色游戏

极客学院在线视频学习网址：
http://www.jikexueyuan.com/course/167_3.html
手机扫描二维码

看你有多色-游戏制作

本节学习绘制看你有多色游戏 UI 并且完成游戏的设计。

1. 增加主要的 `<script src="rect.js"></script>` 代码，修改 `<script src="app.js"></script>` 来绘制方块效果。

2. 在 rect.js 中添加如下代码。

代码清单 5-17

```
function Rect(n,color,Rectcolor){
    createjs.Shape.call(this);
    this.setRectType = function(type){
```

```
                this._RectType = type;
                switch (type){
                    case 1:
                        this.setColor(color);
                        break;
                    case 2:
                        this.setColor(Rectcolor);
                        break;
                }
            }
            this.setColor = function(colorString){
                this.graphics.beginFill(colorString);
                this.graphics.drawRect(0,0,getSize()/n-2,getSize()/n-2);
                this.graphics.endFill();
            }
            this.getRectType = function(){
                return this._RectType;
            }
            this.setRectType(1);
        }
        Rect.prototype = new createjs.Shape();
```

Rect(n,color,Rectcolor)函数用面向对象的方法来定义 rect 类的一个实体，传入的参数 1 为方块的个数，参数 2 为默认方块颜色，参数 3 为点击的方块颜色；createjs.Shape.call(this)回调一下这个方法，setRectType 方法设置方块颜色，通过类型设定当前方块颜色，当类型为 1 时，颜色为传入的 color，类型为 2 时，先设置为固定值：Rectcolor；setColor 方法设置方块的方法；beginFill(colorString)设置方块颜色；this.graphics.drawRect(0,0,getSize()/n-2,getSize()/n-2)设置方块的大小；getRectType 方法返回方块的类型；this.setRectType(1)默认的类型为 1；Rect.prototype = new createjs.Shape()初始化 rect 函数。

3. 在 addRect()中添加如下代码。

代码清单 5-18

```
var stage = new createjs.Stage("gameView");
createjs.Ticker.setFPS(30);
createjs.Ticker.addEventListener("tick", stage);
var gameView = new createjs.Container();
stage.addChild(gameView);
function startGame(){
    getCanvasSize();
    n = 2;
    addRect();
}
function addRect() {
    var c1 = parseInt(Math.random() *1000000);
    var color = ("#"+c1);
    var x = parseInt(Math.random() * n);
    var y = parseInt(Math.random() * n);
    for (var indexX = 0; indexX < n; indexX++) {
        for (var indexY = 0; indexY < n; indexY++) {
            var c2 = parseInt(c1-10*(n-indexY));
            var   Rectcolor = ("#"+c2);
```

```
                    var r = new Rect(n,color,Rectcolor);
                    gameView.addChild(r);
                    r.x = indexX;
                    r.y = indexY;
                    if (r.x == x && r.y == y) {
                        r.setRectType(2);
                    }
                    r.x = indexX * (getSize() / n);
                    r.y = indexY * (getSize() / n);
                    if (r.getRectType() == 2) {
                        r.addEventListener("click",clickRect)
                    }        }      }
}
function clickRect() {
    if (n < 7) {
        ++n;
    }
    gameView.removeAllChildren();
    addRect();
}
function getCanvasSize() {
    var gView = document.getElementById("gameView");
    gView.height = window.innerHeight – 4;
    gView.width = window.innerWidth – 4;
}
function getSize() {
    if (window.innerHeight >= window.innerWidth) {
        return window.innerWidth;
    } else {
        return window.innerHeight;
    }}
startGame();
```

4. 函数的入口为 startGame()，该函数调用 getCanvasSize() 获取 Canvas 的大小；然后默认方格为 2×2 行，接着调用 addRect() 函数添加方格。

5. addRect() 先设定方格的颜色 color 为一个随机值，设定随机的 x/y 轴坐标，然后使用一个 for (var indexX = 0; indexX < n; indexX++) {for (var indexY = 0; indexY < n; indexY++)} 循环来绘制每个方格，当方格的 x/y 轴正好等于设定的 x/y 轴坐标时，用 setRectType(2) 设定其类型为 2，这样重新绘制该方格颜色为 Rectcolor 值，目前该代码的 Rectcolor 取值还不够科学，学员可以作为加强练习继续优化；最后当类型为 2 时，用 clickRect() 函数响应点击事件。

6. clickRect() 函数判断方格时为最大 7 个（如果不是继续添加方格），接着清除所有子元素，重新调用 addRect() 函数绘制方格。

7. 添加网页的 CSS 效果。

代码清单 5-19

```
*{
    margin: 0px;
    padding: 0px;
}
```

```
.main{
    width: 80%;
    margin: 0px 2px;
}
#gameView{
    width: 100%;
    margin: 20px auto;
}
```

8. 到此为止看你有多色游戏已经基本完成，运行之后，刚载入页面时效果如图 5-20 所示。

图 5-20　看你有多色游戏界面

点击第一个图片后方格增加为 3×3 个，并且颜色更难区分，如图 5-21 所示。

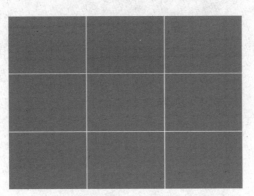

图 5-21　看你有多色游戏 3×3 方格界面

如图 5-22 所示，直至方格成为 4×4 后不再增加方格，而是让颜色相似度更接近。

图 5-22　看你有多色游戏最后显示界面效果

第6章

HTML5大型游戏——太空英雄大战

■ 本章是将所学的知识在一个大型游戏项目中进行综合应用,可系统地复习前面章节所学的知识,并且可以熟悉 HTML5 大型游戏开发的基本流程。

6.1 游戏简介

太空英雄大战游戏是一个大型的射击游戏，如图 6-1 所示，敌机从屏幕上方飞落下来，在飞落过程中不断发射子弹，英雄飞船射击子弹去攻击敌机，可以使用键盘的方向键控制英雄飞船的移动位置。

在进入游戏编码之前，先来浏览一下需要在项目中使用的文件，并对相关文件进行布局。

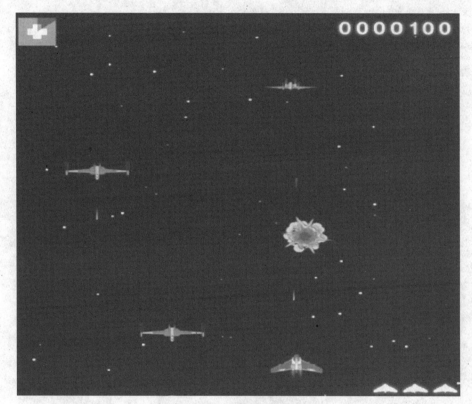

图 6-1　太空英雄大战全景图

6.2 准备项目

项目需要很多文件，文件的目录配置和开发的流程都需要设计，本节主要介绍项目文件和准备工作。

6.2.1 设置 HTML 文件

项目使用的 HTML 文件里包含了很多 JavaScript 的脚本文件，也创建了应用程序的实例，并且定义了一个 init()初始化函数，索引文件用来定义在项目中使用到的参数列表。代码清单 6-1 是 index.html 文件。

代码清单 6-1

```
<!DOCTYPE html>
<html>
<head>
```

```html
<title>SPACE HERE</title>
<!--CREATEJS-->
<script src="js/lib/easeljs-0.7.1.min.js"></script>
<script src="js/lib/soundjs-0.5.2.min.js"></script>
<script src="js/lib/preloadjs-0.4.1.min.js"></script>
<script src="js/lib/tweenjs-0.5.1.min.js"></script>
<script src="js/lib/BitmapText.js"></script>
<!--NDGMR-->
<script src="js/lib/ndgmr.Collision.sprite.js"></script>
<!--GAME CLASSES-->
<script src="js/state.js"></script>
<script src="js/utils/Utils.js"></script>
<!--COMPONENTS-->
<script src="js/classes/components/Preloader.js"></script>
<script src="js/classes/components/SimpleButton.js"></script>
<!--SCENES-->
<script src="js/scenes/GameMenu.js"></script>
<script src="js/scenes/Game.js"></script>
<script src="js/scenes/GameOver.js"></script>
<!--MANAGERS-->
<script src="js/classes/managers/AssetManager.js"></script>
<script src="js/classes/managers/SpritePool.js"></script>
<!--SPRITES-->
<script src="js/classes/sprites/Explosion.js"></script>
<script src="js/classes/sprites/HealthMeter.js"></script>
<script src="js/classes/sprites/LifeBox.js"></script>
<script src="js/classes/sprites/Scoreboard.js"></script>
<script src="js/classes/sprites/EnemyShip.js"></script>
<script src="js/classes/sprites/HeroShip.js"></script>
<script src="js/classes/sprites/Bullet.js"></script>
<!--APPLICATION-->
<script src="js/SpaceHero.js"></script>
</head>
<body onload="init();" style="background-color: black">
<img src="img/bg.png" style="position: absolute">
<canvas id="canvas" width="600" height="700" style="position: absolute"></canvas>
</body>
<script>
    var ARROW_KEY_SPACE = 32;
    var ARROW_KEY_UP = 38;
    var ARROW_KEY_DOWN = 40;
    var ARROW_KEY_LEFT = 37;
    var ARROW_KEY_RIGHT = 39;
    var stage;
    var canvas;
    var spritesheet;
    var screen_width;
    var screen_height;
    function init() {
        window.game = window.game || {};
        game.main = new game.SpaceHero();
```

```
    }
</script>
</html>
```

如代码清单 6-1 所示，index.html 文件中包含非常多的 JavaScript 脚本文件，下面将详细说明这些文件的使用。

6.2.2 Sprite Sheet 文件

本章的 Sprite Sheet 文件在源码目录中，使用 Flash CC 来制作动画帧效果（可参照相关技术文档），如图 6-2 所示，sprites.png 文件包含了游戏使用的大部分动画和静止帧的图像效果（文件的背景图像是由 Canvas DOM 元素加载的）。

图 6-2　太空英雄大战 Sprite Sheet 图

该图片中包含了多数在游戏中使用到的动画帧，包括子弹爆炸效果，甚至为扩展游戏而预留的一些图像效果。代码清单 6-2 所示是：太空英雄大战的 Sprite Sheet 效果队列保存在 JSON 对象中，当预加载时加入到项目中。

代码清单 6-2

```
{
"framerate":24,
"images":["assets/all.png"],
"frames":[
    ......
],
"animations":{
    "1": {"frames": [138], "speed": 1},
    "2": {"frames": [139], "speed": 1},
    "4": {"frames": [141], "speed": 1},
    "5": {"frames": [142], "speed": 1},
    "6": {"frames": [143], "speed": 1},
    "7": {"frames": [144], "speed": 1},
    "0": {"frames": [137], "speed": 1},
    "8": {"frames": [145], "speed": 1},
    "3": {"frames": [140], "speed": 1},
    "9": {"frames": [146], "speed": 1},
    "asteroid3": {"frames": [37], "speed": 1},
```

```
"enemy2Hit": {
    "frames": [21, 22, 23, 24, 25, 26],
    "next": "enemy2Idle",
    "speed": 1
},
"powerup": {
    "frames": [39, 40, 41, 42,... 39 ],
    "speed": 1
},
"asteroid4": {"frames": [38], "speed": 1},
"shield": {
    "frames": [68, 69, 70,...68 ],
    "speed": 1
},
"life": {
    "frames": [147, 148, 149, 150, 151, 152, 153, 154, 155, 98],
    "speed": 0.4
},
"shieldHUD": {"frames": [97], "speed": 1},
"heroPrize": {
    "frames": [1, 2, 3, 4, 5, 6, 7, 8, 9, 10, 11, 12, 13, 0],
    "speed": 1
},
"progessHUD": {
    "frames": [98, 99, 99,...133],
    "speed": 1
},
"powerHUD": {"frames": [134], "speed": 1},
"enemy1Hit": {
    "frames": [28, 29, 30, 31, 32, 33],
    "next": "enemy1Idle",
    "speed": 1
},
"asteroid2": {"frames": [36], "speed": 1},
"enemy1Idle": {"frames": [27], "speed": 1},
"healthHUD": {"frames": [34], "speed": 1},
"star1": {"frames": [98], "speed": 1},
"explosion": {
    "frames": [156, 157, 158,...173],
    "speed": 0.4
},
"star2": {"frames": [174], "speed": 1},
"star3": {"frames": [175], "speed": 1},
"heroIdle": {"frames": [0], "speed": 1},
"enemy2Idle": {"frames": [20], "speed": 1},
"gameOver": {"frames": [177], "speed": 1},
"asteroid1": {"frames": [35], "speed": 1},
"bullet": {"frames": [135, 136], "speed": 1},
"heroHit": {
    "frames": [0, 14, 15, 16, 17, 18, 19],
    "next": "heroIdle",
```

```
            "speed": 0.4
        },
        "title": {"frames": [176], "speed": 1}
    }
}
```

如上所示，定义了 speed 和 next 属性来展示运行帧效果和下一帧加载的图像。

6.2.3 资源管理

资源管理在文件 AssetManager.js 中，该文件定义需要加载的声音、图像、数据、事件等参数，这些参数是在初始化 init()函数时进行定义和加载的。

代码清单 6-3

```
//sounds
p.EXPLOSION = 'explosion';
p.SOUNDTRACK = 'soundtrack';

//graphics
p.GAME_SPRITES = 'game sprites';

//data
p.GAME_SPRITES_DATA = 'game sprites data'

//events
p.ASSETS_PROGRESS = 'assets progress';
p.ASSETS_COMPLETE = 'assets complete';

p.assetsPath = 'assets/';

p.loadManifest = null;
p.queue = null;
p.loadProgress = 0;


p.initialize = function () {
    this.EventDispatcher_initialize();
    this.loadManifest = [
        {id:this.EXPLOSION, src:this.assetsPath + 'explosion.mp3'},
        {id:this.SOUNDTRACK, src:this.assetsPath + 'dreamRaid1.mp3'},
        {id:this.GAME_SPRITES_DATA, src:this.assetsPath + 'all.json'},
        {id:this.GAME_SPRITES, src:this.assetsPath + 'all.png'}
    ];
}
```

6.2.4 创建应用类

SpaceHero.js 文件用来加载游戏的主要程序框架。

代码清单 6-4

```
(function (window) {

    window.game = window.game || {}
```

```javascript
function SpaceHero() {
    this.initialize();
}

var p = SpaceHero.prototype;

p.preloader;

p.currentGameStateFunction;
p.currentScene;


p.initialize = function () {
    // set globals
    canvas = document.getElementById('canvas');
    stage = new createjs.Stage(canvas);
    screen_width = canvas.width;
    screen_height = canvas.height;
    // end globals
    createjs.Touch.enable(stage);
    stage.enableMouseOver();
    game.assets = new game.AssetManager();
    this.preloadAssets()
}
p.preloadAssets = function () {
    this.preloader = new ui.Preloader('#d2354c', '#FFF');
    this.preloader.x = (canvas.width / 2) - (this.preloader.width / 2);
    this.preloader.y = (canvas.height / 2) - (this.preloader.height / 2);
    stage.addChild(this.preloader);
    game.assets.on(game.assets.ASSETS_PROGRESS, this.onAssetsProgress, this);
    game.assets.on(game.assets.ASSETS_COMPLETE, this.assetsReady, this);
    game.assets.preloadAssets();
}
p.onAssetsProgress = function () {
    this.preloader.update(game.assets.loadProgress);
    stage.update();
}
p.assetsReady = function () {
    stage.removeChild(this.preloader);
    stage.update();
    this.createSpriteSheet();
    this.gameReady();
}
p.createSpriteSheet = function () {
    var assets = game.assets;
    spritesheet = new createjs.SpriteSheet(assets.getAsset(assets.GAME_SPRITES_DATA));
}
p.gameReady = function () {
    createjs.Ticker.setFPS(60);
    createjs.Ticker.on("tick", this.onTick, this);
```

```javascript
            this.changeState(game.GameStates.MAIN_MENU);
        }
        p.changeState = function (state) {
            switch (state) {
                case game.GameStates.MAIN_MENU:
                    this.currentGameStateFunction = this.gameStateMainMenu;
                    break;
                case game.GameStates.GAME:
                    this.currentGameStateFunction = this.gameStateGame;
                    break;
                case game.GameStates.RUN_SCENE:
                    this.currentGameStateFunction = this.gameStateRunScene;
                    break;
                case game.GameStates.GAME_OVER:
                    this.currentGameStateFunction = this.gameStateGameOver;
                    break;
            }
        }
        p.onStateEvent = function (e, obj) {
            this.changeState(obj.state);
        }
        p.disposeCurrentScene = function () {
            if (this.currentScene != null) {
                stage.removeChild(this.currentScene);
                if(this.currentScene.dispose){
                    // this.currentScene.dispose();
                }
                this.currentScene = null;
            }
        }
        p.gameStateMainMenu = function (tickEvent) {
            var scene = new game.GameMenu();
            scene.on(game.GameStateEvents.GAME, this.onStateEvent, this, true, {state:game.GameStates.GAME});
            stage.addChild(scene);
            this.disposeCurrentScene();
            this.currentScene = scene;
            this.changeState(game.GameStates.RUN_SCENE);
        }
        p.gameStateGame = function (tickEvent) {
            var scene = new game.Game();
            scene.on(game.GameStateEvents.GAME_OVER, this.onStateEvent, this, true, {state:game.GameStates.GAME_OVER});
            stage.addChild(scene);
            this.disposeCurrentScene()
            this.currentScene = scene;
            this.changeState(game.GameStates.RUN_SCENE);
        }
        p.gameStateGameOver = function (tickEvent) {
            var scene = new game.GameOver();
            stage.addChild(scene);
```

```
            scene.on(game.GameStateEvents.MAIN_MENU, this.onStateEvent, this, true, {state:game.Game
States.MAIN_MENU});
            scene.on(game.GameStateEvents.GAME, this.onStateEvent, this, true, {state:game.Game
States.GAME});
            this.disposeCurrentScene();
            this.currentScene = scene;
            this.changeState(game.GameStates.RUN_SCENE);
        }
        p.gameStateRunScene = function (tickEvent) {
            if (this.currentScene.run) {
                this.currentScene.run(tickEvent);
            }
        }
        p.onTick = function (e) {
            if (this.currentGameStateFunction != null) {
                this.currentGameStateFunction(e);
            }
            stage.update();
        }

        window.game.SpaceHero = SpaceHero;

}(window));
```

6.3 创建 Sprites

太空英雄大战游戏中的 Spite 主要包括一个英雄飞船、两个敌方飞船、子弹和爆炸效果,本节开始准备这些游戏中使用到的 Sprites。

6.3.1 创建英雄飞船

英雄飞船是由玩家进行控制的,驾驶的飞船需要躲避敌机发射的炸弹并且自己发射炸弹来击毁无数的地方飞船,图 6-3 是英雄飞船的帧效果图。

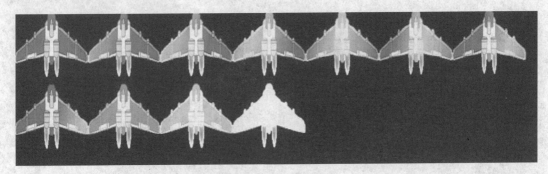

图 6-3 太空英雄动画帧

如图 6-3 所示,这里有三个英雄飞船的动画帧,前两个为 idle 帧和 damage 顺序帧,damage 帧主要表示当受到敌方飞船炸弹攻击时英雄飞船逐渐变白的效果,第三个动画帧设计为当能量加强时英雄飞船颜色逐渐加深的效果。代码清单 6-5 显示了在 HeroShip.js 创建和使用 idle 帧和 damage 帧的定义。

代码清单 6-5

```javascript
(function (window) {

    window.game = window.game || {};

    function HeroShip() {
        this.initialize();
    }

    var p = HeroShip.prototype = new createjs.Sprite();

    p.Sprite_initialize = p.initialize;

    p.EXPLOSION_COMPLETE = 'explosion complete';
    p.EXPLOSION_OFFSET = 55;
    p.INVINCIBLE_TIME = 1500;

    p.invincible = false;
    p.shouldDie = false;
    p.speed = 500;
    p.nextX = null;
    p.nextY = null;

    p.initialize = function () {
        this.Sprite_initialize(spritesheet, "heroIdle");
        this.regX = this.getBounds().width / 2;
        this.regY = this.getBounds().height / 2;
    }
    p.takeDamage = function () {
        this.gotoAndPlay("heroHit");
    }
    p.explode = function () {
        this.gotoAndPlay('explosion');
        this.regX = this.regY = this.EXPLOSION_OFFSET;
        this.on('animationend', this.explosionComplete, this, true);
        createjs.Sound.play(game.assets.EXPLOSION);
    }
    p.explosionComplete = function (e) {
        this.stop();
        this.dispatchEvent(this.EXPLOSION_COMPLETE);
    }
    p.reset = function () {
        this.shouldDie = false;
        this.gotoAndStop('heroIdle');
        this.regX = this.getBounds().width / 2;
        this.regY = this.getBounds().height / 2;
    }
    p.makeInvincible = function () {
        this.invincible = true;
        this.alpha = .4;
        setTimeout(this.removeInvincible.bind(this), this.INVINCIBLE_TIME);
```

```
        }
        p.removeInvincible = function () {
            this.invincible = false;
            this.alpha = 1;
        }

        window.game.HeroShip = HeroShip;

}(window));
```

函数开始定义一些参数，如爆炸事件参数、爆炸偏移参数、英雄飞船新生命的默认存活时间，接着两个布尔函数来定义英雄飞船是否无限存活和是否在下一次渲染的时候死亡，然后定义了两个位移参数。

takeDamage 方法调用英雄飞船炸弹碰撞效果，使用 heroHit 序列来产生白色的亮光；explode 方法调用循环的渲染效果来不断地播放声音和帧变化，固定爆炸的坐标，产生一个名为 explosionComplete 的监听事件来停止动画效果和分发 EXPLOSION_COMPLETE 事件。

reset 方法重设 shouldDie 属性为 false，将当前帧设定为 idle，并重设坐标。最后 makeInvincible 方法通过设置 invincible 参数为 true、设置 alpha 为 .4 来恢复英雄飞船，通过在 setTimeout 方法中调用 INVINCIBLE 参数来重设无敌能力。

6.3.2 创建敌方飞船

游戏中有两种敌机，每个敌机有其自己的爆破动画帧效果，敌机不断发射炸弹并往下移动，如图 6-4 是敌机飞船的图片效果。

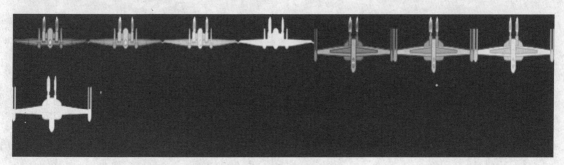

图 6-4 敌机飞船动画帧

如代码清单 6-6 所示，虽然有两种敌机，但只创建一个 Sprite 类，飞船的效果图是 1 还是 2 是系统自由选择的，每种飞船有自己的一系列帧设置来显示。

代码清单 6-6

```
(function (window) {
    window.game = window.game || {}

    function EnemyShip(startX) {
        this.initialize(startX);
    }
    var p = EnemyShip.prototype = new createjs.Sprite();
    p.Sprite_initialize = p.initialize;
    p.type = null;
    p.HP = null;
    p.points = null;
```

```
        p.lastFired = 0;
        p.fireDelay = 2000;

        p.speed = 150;
        p.nextY = 0;
        p.shouldDie = false;

        p.initialize = function (startX) {
            this.type = Utils.getRandomNumber(0, 2) + 1;
            this.HP = this.type * 3;
            this.points = this.type * 100;
            this.Sprite_initialize(spritesheet, "enemy" + this.type + "Idle");
            this.regX = this.getBounds().width / 2;
            this.regY = this.getBounds().height / 2;

        }
        p.takeDamage = function () {
            this.gotoAndPlay("enemy" + this.type + "Hit");
            this.HP--;
            if (this.HP <= 0) {
                this.shouldDie = true;
            }
        }
        p.reset = function () {
            this.type = Utils.getRandomNumber(0, 2) + 1;
            this.shouldDie = false;
            this.HP = this.type * 3;
            this.points = this.type * 100;
            this.gotoAndPlay("enemy" + this.type + "Idle");
        }

        window.game.EnemyShip = EnemyShip;

}(window));
```

　　type、HP 和 points 属性是每个飞船的重要属性。这些属性是在 init 函数初始化过程中随机定义的值，它用来制定框架结构、袭击的坐标和每个飞船的 sheet 值，下一个值是用来定义飞行中子弹的信息；lastFired 参数用来定义上一个敌方子弹的发射时间，firedelay 用来定义自从上次攻击到现在的时间；speed 用来定义敌机子弹降落屏幕舞台的速度；nextY 在游戏循环过程中定义更新/渲染时使用，shouldDie 定义在下一次渲染过程中飞船是否爆炸的属性。

　　接下去是对一系列方法的定义，initialize 方法用来随机选择一种敌机飞船并分配相应的资源；当飞船被碰到的时候，takeDamage 方法将被立即调用其飞船种类相应的 enemy1Hit 或者 enemy2Hit 函数，HP 参数自减，当 HP 参数≤0 时，设置 ShouldDie 参数值为 true；这个属性将在游戏循环的渲染环节被调用。

　　最后 reset 方法重新设置飞船类型，加载飞船属性，这种加载技术将在使用对象池章节进行更为详细的介绍。

6.3.3　创建子弹和爆炸效果

　　敌机飞船和英雄飞船都发射子弹，子弹有一定运行速度且方法决定于发射飞船的方向，当子弹碰到飞船时，爆炸帧将代替飞船的位置，图 6-5 所示是炸弹和爆炸效果的图片帧。其中有两种炸弹帧，一种红色，另一种绿色，红色的属于英雄飞船发射，绿色属于敌机飞船发射，代码清单 6-7 构建了相应的帧效果。

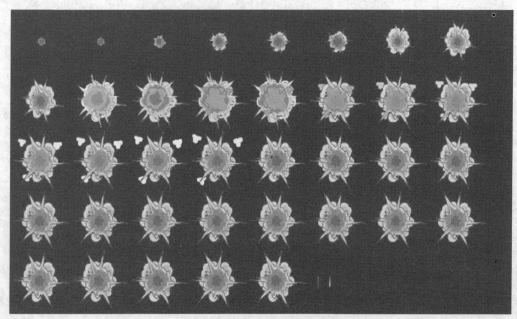

图 6-5　炸弹和爆炸效果帧

代码清单 6-7

```
(function (window) {
    window.game = window.game || {}
    function Bullet() {
        this.initialize();
    }
    var p = Bullet.prototype = new createjs.Sprite();
    p.Sprite_initialize = p.initialize;
    p.speed = 500;
    p.nextY = null;
    p.shouldDie = false;
    p.initialize = function () {
        this.Sprite_initialize(spritesheet, "bullet");
        this.paused = true;
    }
    p.reset = function () {
        this.shouldDie = false;
    }
    window.game.Bullet = Bullet;
}(window));
```

与 enemyShip 类相同，子弹 Sprite 也有其速度和下一个运行位移，外加一个布尔值来指定其下个渲染循环的生死属性，子弹 Sprite 有两组帧，在帧 1 中初始化为英雄子弹的红色效果，帧 2 中定义为敌机子弹效果，创建对象池来保存这些子弹，所以需要一个 reset 方法来重置这些下一次使用的对象。代码清单 6-8 中，爆炸类的设置也与此类似。

代码清单 6-8

```
(function (window) {
    window.game = window.game || {}
    function Explosion() {
        this.initialize();
    }
    var p = Explosion.prototype = new createjs.Sprite();
    p.Sprite_initialize = p.initialize;
    p.initialize = function () {
        this.Sprite_initialize(spritesheet,'explosion');
        this.paused = true;
    }
    p.reset = function(){
        this.gotoAndStop('explosion');
    }
    window.game.Explosion = Explosion;
}(window));
```

6.4 创建参谋中心（HUD）

HUD 是参谋中心（heads-up display）的首字缩写，参谋中心保存游戏等级信息，在太空英雄大战游戏中一共有三个 HUD 元素：健康测量中心、分数和生命指示器，这三个 HUD 元素都是 Sprites 完成的。

6.4.1 创建 HUD Sprite 框架

图 6-6 所示是 HUD 游戏元素的一系列帧和静止状态图片，前面一部分奇怪的图形是健康测量帧图片用时针帧效果来显示英雄的摧毁状态，它将与右下角的+图形一起表示不同的健康状态效果；飞船突变用来表示还有几条生命，它包括一系列帧来制定飞船的爆炸消失效果，最下面的数字 0～9 用来显示分数的位图。

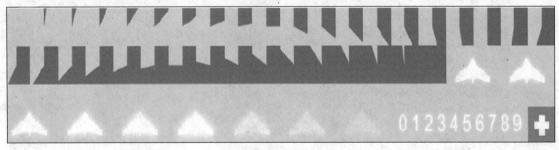

图 6-6　HUD 中心帧

6.4.2 创建 HUD 效果图

健康测量显示当前飞船生命的状态，由一个固定的 Sprite 作为底色加一个变化的 Sprite 帧来显示，如图 6-7 所示。

HealthMeter 类扩展 container 并且控制健康监测效果。

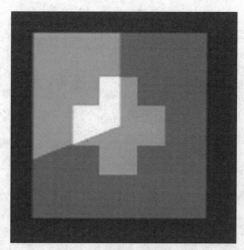

图 6-7 健康进度条显示效果

代码清单 6-9

```
(function (window) {
    window.game = window.game || {}
    function HealthMeter() {
        this.initialize();
    }
    var p = HealthMeter.prototype = new createjs.Container();
    p.meter = null;
    p.maxDamage = null;
    p.damage = 0;
    p.empty = false;
    p.Container_initialize = p.initialize;
    p.initialize = function () {
        this.Container_initialize();
        this.x = this.y = 5;
        this.buildMeter();
    }
    p.buildMeter = function () {
        var health = new createjs.Sprite(spritesheet, 'healthHUD');
        this.meter = new createjs.Sprite(spritesheet, 'progessHUD');
        this.maxDamage =
this.meter.spriteSheet.getAnimation(this.meter.currentAnimation).frames.length - 1;
        this.meter.paused = true;
        this.addChild(health, this.meter);
    }
    p.takeDamage = function (damage) {
        this.damage += damage;
        var perc = this.damage / this.maxDamage > 1 ? 1 : this.damage / this.maxDamage;
        var frame = (this.maxDamage * perc);
        createjs.Tween.get(this.meter).to({currentAnimationFrame:frame}, 100)
            .call(this.checkHealth, null, this);
    }
    p.checkHealth = function (e) {
        if (this.meter.currentAnimationFrame === this.maxDamage) {
```

```
                this.empty = true;
            }
        }
        p.reset = function (e) {
            this.damage = 0;
            this.empty = false;
            this.meter.currentAnimationFrame = 0;
        }
        window.game.HealthMeter = HealthMeter;

}(window));
```

Meter 属性保存在英雄飞船损坏时的测量帧的信息，maxdamage 属性值在 buildMeter 方法进行设定匹配最大的策略帧，damage 在英雄飞船被攻击自增，并且保存下一个测量帧序列；empty 属性在到达最后的帧时设置为 true，并告诉游戏飞船将要爆炸了；buildMeter 方法用于定义健康中心的设置，用 addChild 将该 Sprite 添加到 Container 中去。

takeDamage 方法在子弹碰到英雄飞船的时候被调用，传入 damage 参数实现可写性，用公式来计算飞船的损坏程度，最后，在目标中使用 TweenJS 动态方法来设定此时的动画帧，如果所有的测量帧全都使用完毕、静止状态的健康 Sprite 已经全部被测量帧覆盖了，则调用 checkHealth 函数来设置 empty 属性为 true；最后 reset 函数在玩家重新开始飞船生命时重设测量值。

6.4.3 创建分数板

分数板使用位图格式来显示游戏的分数状态，代码清单 6-10 是 Scoreboard 类，定义了一个 Container、当分数更新时在 Container 上添加一个 BitmapText 对象。

代码清单 6-10

```
(function (window) {
    window.game = window.game || {}
    function Scoreboard() {
        this.initialize();
    }
    var p = Scoreboard.prototype = new createjs.Container();
    p.scoreTxt;
    p.score = 0;
    p.Container_initialize = p.initialize;
    p.initialize = function () {
        this.Container_initialize();
        this.x = screen_width - 165;
        this.y = 5;
        this.updateScore(0);
    }
    p.updateScore = function (points) {
        var formattedScore;
        this.removeAllChildren();
        this.score += points;
        formattedScore = this.addLeadingZeros(this.score, 7);
        this.scoreTxt = new createjs.BitmapText(formattedScore, spritesheet);
        this.addChild(this.scoreTxt);
    }
    p.addLeadingZeros = function (score, width) {
```

```
                score = score + '';
                return score.length >= width ? score : new Array(width - score.length + 1).join(0) + score;
            }
            p.getScore = function () {
                return this.addLeadingZeros(this.score, 7);
            }
            window.game.Scoreboard = Scoreboard;
        }(window));
```

Scoreboard 对象有两个属性值：分数和分数显示 BitmapText 对象，当分数板更新时，updateScore 方法被调用来更新现在的分数，用 this.removeAllChildren();函数来清除 Container 上的所有子层，然后再用 addLeadingZeros()来设定分值显示格式为 7 位，不足的位置补上 0（如图 6-8 所示），最后调用新建 BitmapText 的方法添加相应文本位图。getScore 方法用来得到游戏的分数，这个值在游戏结束获取分数时调用。

图 6-8　分数板显示效果

6.4.4　创建生命箱

设计生命箱主要用来显示玩家还剩多少条飞船的生命，生命箱用相应的 Sprite 进行加载并显示在游戏区域的右下角，LifeBox 类的代码如下所示。

代码清单 6-11

```
(function (window) {
    window.game = window.game || {};
    function LifeBox(numLives) {
        this.numLives = numLives;
        this.initialize();
    }
    var p = LifeBox.prototype = new createjs.Container();
    p.numLives = null;
    p.Container_initialize = p.initialize;
    p.initialize = function () {
        this.Container_initialize();
        this.buildSprites();
        this.positionBox();
    }
    p.buildSprites = function () {
        var i, life;
        var xPos = 0;
        for (i = 0; i < this.numLives; i++) {
            life = new createjs.Sprite(spritesheet, 'life');
            life.paused = true;
            life.x = xPos;
            this.addChild(life);
```

```
                xPos += life.getBounds().width;
            }
        }
        p.positionBox = function () {
            this.x = screen_width - this.getBounds().width;
            this.y = screen_height - this.getBounds().height;
        }
        p.removeLife = function () {
            var life = this.getChildAt(0);
            life.on('animationend', function (e) {
                e.target.stop();
                this.removeChild(e.target);
            }, this)
            life.play();
        }
        window.game.LifeBox = LifeBox;
}(window));
```

设置一个 Container 容器来装载玩家所拥有生命的 Sprite 列表，通过 LifeBox(numLives)来获取该类的唯一属性值；initialize 方法通过 buildSprites()和 positionBox()函数来创建 Sprites 接着定位它们；buildSprites()中使用 for (i = 0; i < this.numLives; i++)循环串接并显示每个飞船 Sprite；positionBox()定位每个飞船的位置为游戏界面右下角；当玩家失去一条飞船生命时，调用 removeLife 方法来从左移出玩家飞船。图 6-9 所示是 LifeBox 容器中的三个飞船。

图 6-9 英雄飞船列表

6.5 创建对象池

创建一个对象数组来保存一系列类似对象，这个技术叫对象池。

代码清单 6-12

```
(function () {
    var SpritePool = function (type, length) {
        this.pool = [];
        var i = length;
        while (--i > -1) {
            this.pool[i] = new type();
        }
    }
    SpritePool.prototype.getSprite = function () {
        if (this.pool.length > 0) {
            return this.pool.pop();
        }
        else {
            throw new Error("You ran out of sprites!");
```

```
        }
    }
    SpritePool.prototype.returnSprite = function (sprite) {
        this.pool.push(sprite);
    }
    window.game.SpritePool = SpritePool;
})();
```

SpritePool 构造函数包含两个参数：对象类型和对象长度，例如可以使用 var bullets = new game.SpriePool(game.Bullet, 20)来创建 SpritePool 对象，接着使用 getSprite 方法来进行对象实例化：

Var bullet = bullets.getSprite();
stage.addChild(bullet);

当对象使用完毕，用 returnSprite 方法释放对象值以便重复使用：

Bullets.returnSprite(bullet);
Stage.removeChild(bullet);

在该游戏中，可以使用该类建立多数的对象池。

6.6 创建场景

太空英雄大战包含有三个场景，这一节将一一详细介绍，先从菜单场景开始。

6.6.1 创建游戏菜单场景

主菜单从上直下落入屏幕，显示游戏菜单内容，点击游戏按钮即可进入游戏画面，代码清单如下。

代码清单 6-13

```
(function (window) {
    window.game = window.game || {}
    function GameMenu() {
        this.initialize();
    }
    var p = GameMenu.prototype = new createjs.Container();
    p.playBtn;
    p.Container_initialize = p.initialize;
    p.initialize = function () {
        this.Container_initialize();
        this.addTitle();
        this.addButton();
    }
    p.addTitle = function () {
        var titleYPos = 200;
        var title = new createjs.Sprite(spritesheet, 'title');
        title.regX = title.getBounds().width / 2;
        title.x = screen_width / 2;
        title.y = -50;
        createjs.Tween.get(title).to({y: titleYPos}, 5000)
            .call(this.bringTitle, null, this);
        this.addChild(title);
    }
    p.addButton = function () {
        this.playBtn = new ui.SimpleButton('Play Game');
```

```
            this.playBtn.on('click', this.playGame, this);
            this.playBtn.regX = this.playBtn.width / 2;
            this.playBtn.x = canvas.width / 2;
            this.playBtn.y = 400;
            this.playBtn.alpha = 0;
            this.playBtn.setButton({upColor: '#d2354c', color: '#FFF', borderColor: '#FFF', overColor: '#900'});
            this.addChild(this.playBtn);
        }
        p.bringTitle = function (e) {
            createjs.Tween.get(this.playBtn).to({alpha: 1}, 1000);
        }
        p.playGame = function (e) {
            createjs.Sound.play(game.assets.EXPLOSION);
            this.dispatchEvent(game.GameStateEvents.GAME);
        }
        window.game.GameMenu = GameMenu;
}(window));
```

利用 new createjs.Sprite(spritesheet, 'title')载入标题 Sprite，游戏按钮使用 SimpleButton 控件加入到 Container 容器中，然后使用 bringTitle 函数实现自上而下的淡然效果，开始游戏按钮调用 playGame 来启动爆炸声效，并分配适当的游戏事件来启动游戏，主菜单显示如图 6-10 所示。

图 6-10 太空英雄大战开始游戏界面

6.6.2 创建游戏场景

游戏场景用来实现玩游戏的各种绚丽场景，这里包含了游戏的各种主要代码，先了解一下主要的游戏框架，在后面章节再进行详细说明。

代码清单 6-14

```
(function (window) {

    window.game = window.game || {}
```

```
        function Game() {
            this.initialize();
        }
        var p = Game.prototype = new createjs.Container();
        p.Container_initialize = p.initialize;
        p.initialize = function () {
            this.Container_initialize();
    var me = this;
}
        setTimeout = function () {
    me.dispatchEvent(game.GameStateEvents.GAME_OVER);
        }, 3000)
    }
        window.game.Game = Game;
}(window));
```

当游戏初始化时,用 setTimeout 方法来分配游戏事件直至到游戏结束场景,这只是一个游戏框架,在后期有更多的场景容器对象需要进行配置。

6.6.3 创建游戏结束场景

游戏结束场景是在游戏结束时进行显示,GameOver 标题也是来自于 Sprite Sheet,通过 Tween.get(msg).to({scaleX:1, scaleY:1, rotation:360}, 500)将标题旋转放大,addScore()来显示玩家的分数,最后显示两个按钮:一个 playAgain 来重新启动一个游戏,另一个回到主页按钮返回主页菜单。主要的 GameOver 代码如代码清单 6-15 所示。

代码清单 6-15

```
(function (window) {
    window.game = window.game || {}
    function GameOver() {
        this.initialize();
    }
    var p = GameOver.prototype = new createjs.Container();
    p.Container_initialize = p.initialize;
    p.initialize = function () {
        this.Container_initialize();
        createjs.Sound.stop();
        this.addMessage();
        this.addScore();
        this.addButton();
    }
    p.addMessage = function () {
        var msg = new createjs.Sprite(spritesheet, 'gameOver');
        msg.regX = msg.getBounds().width / 2;
        msg.regY = msg.getBounds().height / 2;
        msg.x = screen_width / 2;
        msg.y = 250;
        msg.scaleX = msg.scaleY = 0;
        createjs.Tween.get(msg).to({scaleX:1, scaleY:1, rotation:360}, 500);
        this.addChild(msg);
    }
    p.addScore = function () {
```

```
            var scorePoint = {x:220, y:310};
            var scoreTxt = new createjs.BitmapText(game.score, spritesheet);
            scoreTxt.x = scorePoint.x;
            scoreTxt.y = scorePoint.y;
            this.addChild(scoreTxt);
        }
        p.addButton = function () {
            var playBtn, menuBtn;
            var playBtnPoint = {x:140, y:380};
            var menuBtnPoint = {x:310, y:380};
            var me = this;
            playBtn = new ui.SimpleButton('Play Again');
            playBtn.on('click', this.playAgain, this);
            playBtn.setButton({upColor:'#d2354c', color:'#FFF', borderColor:'#FFF', overColor:'#900'});
            playBtn.x = playBtnPoint.x;
            playBtn.y = playBtnPoint.y;
            this.addChild(playBtn);
            menuBtn = new ui.SimpleButton('Main Menu');
            menuBtn.on('click', this.mainMenu, this);
            menuBtn.setButton({upColor:'#d2354c', color:'#FFF', borderColor:'#FFF', overColor:'#900'});
            menuBtn.x = menuBtnPoint.x;
            menuBtn.y = menuBtnPoint.y;
            this.addChild(menuBtn);
        }
        p.playAgain = function (e) {
            this.dispatchEvent(game.GameStateEvents.GAME);
        }
        p.mainMenu = function (e) {
            this.dispatchEvent(game.GameStateEvents.MAIN_MENU);
        }

        window.game.GameOver = GameOver;

}(window));
```

场景中包含了一系列在屏幕中添加、定位和移动元素的效果，实现的效果如图 6-11 所示。

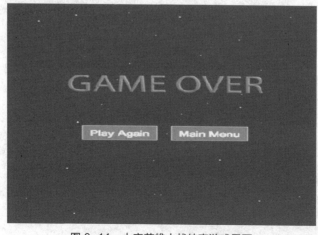

图 6-11 太空英雄大战结束游戏界面

6.7 创建游戏

这一部分将完成游戏设置，使得游戏可以运行起来，在 Game 类中溢出 setTimeout 方法，更改 initialize 方法为如下格式。

代码清单 6-16

```
p.initialize = function () {
        this.Container_initialize();
    var me = this;
}
```

6.7.1 设置游戏参数

游戏中使用很多参数，程序清单 6-17 是对这些参数的分类定义。

代码清单 6-17

```
// Hero
p.heroShip = null;
p.heroBulletPool = null;
p.heroBullets = null;

// Enemies
p.enemyPool = null;
p.enemies = null;
p.enemyBulletPool = null;
p.enemyBullets = null;
p.enemyLastSpawnTime = null;
p.enemySpawnWaiter = 2000;

// SPRITES
p.stars = null;
p.explosionPool = null;
p.healthMeter = null;
p.lifeBox = null;
p.scoreboard = null;

//
p.leftWall = null;
p.rightWall = null;
p.ceiling = null;
p.floor = null;
p.betweenLevels = true;
p.numLives = 3;
p.delta = null;

// Controls
p.leftKeyDown = false;
p.rightKeyDown = false;
p.upKeyDown = false;
p.downKeyDown = false;
```

首先为英雄和敌方设定了一系列参数，这些属性定义了飞船实例、炸弹实例和敌机攻击时的速度属性；接下去定义了一组 Sprites 和 HUD 元素的各类属性值；然后是游戏场景和 Sprites 位移的参数，betweenLevels 是个布尔参数，规定当英雄飞船毁灭时是继续游戏还是暂停游戏；numLives 默认的游戏生命值，delta 在游戏开始时 ticker 中调用并计算移动的 Sprites 的位移值，更为详细的使用可以参照源代码。

6.7.2 初始化游戏

游戏开始时使用 initialize 函数来调用一系列的方法，如代码清单 6-18 所示，第一个方法初始化游戏队列和设置一些游戏属性，然后创建游戏 Sprites、英雄飞船边界和游戏控制方法，最后启动背景音乐；setProperties 方法初始化游戏队列。

代码清单 6-18

```
p.initialize = function () {
    this.Container_initialize();
    this.setProperties();
    this.buildStarField();
    this.buildSprites();
    this.setWalls();
    this.setControls();
    createjs.Sound.play(game.assets.SOUNDTRACK);
}
p.setProperties = function () {
    this.heroBulletPool = [];
    this.heroBullets = [];
    this.enemyPool = [];
    this.enemies = [];
    this.enemyBulletPool = [];
    this.enemyBullets = [];
    this.stars = [];
    this.explosionPool = [];
    this.betweenLevels = false;
    this.enemyLastSpawnTime = 0;
}
```

6.7.3 创建游戏 Sprites

在游戏开始时需要初始化很多 Sprites 对象，createjs.Sprite(spritesheet, 'star3')来添加星空效果，buildSprites 方法中 game.HeroShip()创建英雄飞船对象，SpritePool(game.Bullet, 20)建立英雄对象池，game.HealthMeter()建立 HUD 元素等，更具体的参照代码清单 6-19。

代码清单 6-19

```
p.buildStarField = function () {
    var star;
    var numStars = 20;
    for (i = 0; i < numStars; i++) {
        star = new createjs.Sprite(spritesheet, 'star3');
        star.speed = Utils.getRandomNumber(100, 200);
        star.x = Math.random() * screen_width;
        star.y = Math.random() * screen_height;
```

```
            this.addChild(star);
            this.stars.push(star);
        }
    }
    p.buildSprites = function () {
        this.heroShip = new game.HeroShip();
        this.heroShip.on(this.heroShip.EXPLOSION_COMPLETE, this.checkGame, this);
        this.heroShip.x = screen_width / 2;
        this.heroShip.y = screen_height - this.heroShip.getBounds().height;
        this.heroBulletPool = new game.SpritePool(game.Bullet, 20);
        this.enemyBulletPool = new game.SpritePool(game.Bullet, 20);
        this.enemyPool = new game.SpritePool(game.EnemyShip, 10);
        this.explosionPool = new game.SpritePool(game.Explosion, 10);
        this.healthMeter = new game.HealthMeter();
        this.scoreboard = new game.Scoreboard();
        this.lifeBox = new game.LifeBox(this.numLives);
        this.addChild(this.heroShip, this.healthMeter, this.scoreboard, this.lifeBox);
    }
```

6.7.4 设计游戏控制

如代码清单 6-20 所示，飞船移动的边界由 setWalls 方法进行控制，接着为 document 添加一系列的键盘事件响应恰当的 handler。

代码清单 6-20

```
    p.setWalls = function () {
        this.leftWall = this.heroShip.getBounds().width / 2;
        this.rightWall = screen_width - this.heroShip.getBounds().width / 2;
        this.floor = screen_height - this.heroShip.getBounds().height;
        this.ceiling = screen_height - (this.heroShip.getBounds().height * 3);
    }
    p.setControls = function () {
        document.onkeydown = this.handleKeyDown.bind(this);
        document.onkeyup = this.handleKeyUp.bind(this);


    }
    p.handleKeyDown = function (e) {
        e = !e ? window.event : e;
        switch (e.keyCode) {
            case ARROW_KEY_LEFT:
                this.leftKeyDown = true;
                break;
            case ARROW_KEY_RIGHT:
                this.rightKeyDown = true;
                break;
            case ARROW_KEY_UP:
                this.upKeyDown = true;
                break;
            case ARROW_KEY_DOWN:
                this.downKeyDown = true;
                break;
        }
```

```
}
p.handleKeyUp = function (e) {
    e = !e ? window.event : e;
    switch (e.keyCode) {
        case ARROW_KEY_LEFT:
            this.leftKeyDown = false;
            break;
        case ARROW_KEY_RIGHT:
            this.rightKeyDown = false;
            break;
        case ARROW_KEY_SPACE:
            this.spawnHeroBullet();
            break;
        case ARROW_KEY_UP:
            this.upKeyDown = false;
            break;
        case ARROW_KEY_DOWN:
            this.downKeyDown = false;
            break;
    }
}
```

6.7.5 创建游戏循环

如代码清单 6-21 所示，游戏的循环是通过 run 方法中的 tick 事件引起的，将 tickEvent.delta 赋值给 delta，保存移动 Sprites 更新时的下一个移动位置，如果 betweenLevels 不在暂停状态，则调用一系列更新 update 和渲染 render 方法。

代码清单 6-21

```
p.run = function (tickEvent) {
    this.delta = tickEvent.delta;
    if (!this.betweenLevels) {
        this.update();
        this.render();
        this.checkForEnemySpawn(tickEvent.time);
        this.checkForEnemyFire(tickEvent.time);
        this.checkHeroBullets();
        if (!this.heroShip.invincible) {
            this.checkEnemyBullets();
            this.checkShips();
        }
        this.checkHealth();
        this.checkHero();
    }
}
p.update = function () {
    this.updateStars();
    this.updateHeroShip()
    this.updateEnemies();
    this.updateHeroBullets();
    this.updateEnemyBullets();
```

```
}
p.render = function () {
    this.renderStars();
    this.renderHeroShip();
    this.renderEnemies();
    this.renderHeroBullets();
    this.renderEnemyBullets();
}
```

update 和 render 方法在游戏循环中计算并且移动 Sprites 的位置。

6.7.6 设置游戏更新功能

Update 函数计算活动 Sprites 的当前位移和状态，并决定下一次渲染的方式。

1. 更新星空文件

如代码清单 6-22 所示，所有星星以固定的方式从屏幕上往下落，用一个 for 循环来决定每个星星的移动方式。

代码清单 6-22

```
p.updateStars = function () {
    var i, star, velY, speed, nextY;
    var len = this.stars.length;
    for (i = 0; i < len; i++) {
        star = this.stars[i];
        velY = star.speed * this.delta / 1000;
        nextY = star.y + velY;
        if (nextY > screen_height) {
            nextY = -10;
        }
        star.nextY = nextY;
    }
}
```

2. 更新英雄和敌方飞船

如代码清单 6-23 所示，所有飞船显示方式与星星显示方式类似，都是用 nextY 属性来决定下一个渲染循环效果。

代码清单 6-23

```
p.updateHeroShip = function () {
    var velocity = this.heroShip.speed * this.delta / 1000;
    var nextX = this.heroShip.x;
    var nextY = this.heroShip.y;
    if (this.leftKeyDown) {
        nextX -= velocity;
        if (nextX < this.leftWall) {
            nextX = this.leftWall;
        }
    }
    else if (this.rightKeyDown) {
        nextX += velocity;
        if (nextX > this.rightWall) {
            nextX = this.rightWall;
        }
```

```
        }
        else if (this.downKeyDown) {
            nextY += velocity;
            if (nextY > this.floor) {
                nextY = this.floor;
            }
        }
        else if (this.upKeyDown) {
            nextY -= velocity;
            if (nextY < this.ceiling) {
                nextY = this.ceiling;
            }
        }
        this.heroShip.nextX = nextX;
        this.heroShip.nextY = nextY;
}
p.updateEnemies = function () {
        var enemy, i, velY;
        var len = this.enemies.length - 1;
        for (i = len; i >= 0; i--) {
            enemy = this.enemies[i];
            velY = enemy.speed * this.delta / 1000;
            enemy.nextY = enemy.y + velY;
            if (enemy.nextY > screen_height) {
enemy.reset();
                this.enemyPool.returnSprite(enemy);
                this.removeChild(enemy);
                this.enemies.splice(i, 1);
            }
        }
}
```

updateHeroShip 中使用 velocity 来计算英雄飞船运行速度；相应键盘事件来决定飞船移动的方向，增加了边缘判断事件；

updateEnemies 中敌方飞船的移动是自动的，当敌机飞船穿过屏幕时需要调用 reset 方法来设定一个随机类型和属性的飞船值，然后用 returnSprite(enemy)加入到 enemyPool 中，用 removeChild 从屏幕移除该飞船，并用 enemies.splice(i, 1)从 enemies 队列中删除。

3. 更新英雄和敌方炸弹

炸弹的移动效果类似于飞船，唯一不同的是英雄的炸弹是往上移动的，代码清单 6-24 显示了两类炸弹的更新效果。

代码清单 6-24

```
p.updateHeroBullets = function () {
        var bullet, i, velY;
        var len = this.heroBullets.length - 1;
        for (i = len; i >= 0; i--) {
            bullet = this.heroBullets[i];
            velY = bullet.speed * this.delta / 1000;
            bullet.nextY = bullet.y - velY;
            if (bullet.nextY < 0) {
                this.heroBulletPool.returnSprite(bullet);
```

```
                this.removeChild(bullet);
                this.heroBullets.splice(i, 1);
            }
        }
    }
    p.updateEnemyBullets = function () {
        var bullet, i, velY;
        var len = this.enemyBullets.length - 1;
        for (i = len; i >= 0; i--) {
            bullet = this.enemyBullets[i];
            velY = bullet.speed * this.delta / 1000;
            bullet.nextY = bullet.y + velY;
            if (bullet.nextY > screen_height) {
                this.enemyBulletPool.returnSprite(bullet);
                this.removeChild(bullet);
                this.enemyBullets.splice(i, 1);
            }
        }
    }
```

6.7.7 创建渲染函数

如代码清单 6-25 所示，渲染 Sprites 比较简单，星星、英雄飞船/子弹、敌方飞船/子弹都要随着计算的结果进行渲染至其新的位置。

代码清单 6-25

```
    p.renderStars = function () {
        var i, star;
        for (i = 0; i < this.stars.length; i++) {
            star = this.stars[i];
            star.y = star.nextY;
        }
    }
    p.renderHeroShip = function () {
        this.heroShip.x = this.heroShip.nextX;
        this.heroShip.y = this.heroShip.nextY;
    }
    p.renderHeroBullets = function () {
        var bullet, i;
        var len = this.heroBullets.length - 1;
        for (i = len; i >= 0; i--) {
            bullet = this.heroBullets[i];
            if (bullet.shouldDie) {
                this.removeChild(bullet);
                bullet.reset();
                this.heroBulletPool.returnSprite(bullet);
                this.heroBullets.splice(i, 1);
            }
            else {
                bullet.y = bullet.nextY;
            }
```

```
        }
    }
    p.renderEnemyBullets = function () {
        var bullet, i;
        var len = this.enemyBullets.length - 1;
        for (i = len; i >= 0; i--) {
            bullet = this.enemyBullets[i];
            if (bullet.shouldDie) {
                this.removeChild(bullet);
                bullet.reset();
                this.enemyBulletPool.returnSprite(bullet);
                this.enemyBullets.splice(i, 1);
            }
            else {
                bullet.y = bullet.nextY;
            }
        }
    }
    p.renderEnemies = function () {
        var enemy, i;
        var len = this.enemies.length - 1;
        for (i = len; i >= 0; i--) {
            enemy = this.enemies[i];
            if (enemy.shouldDie) {
                this.scoreboard.updateScore(enemy.points);
                this.enemies.splice(i, 1);
                this.removeChild(enemy);
                this.spawnEnemyExplosion(enemy.x, enemy.y);
                enemy.reset();
                this.enemyPool.returnSprite(enemy);
            }
            else {
                enemy.y = enemy.nextY;
            }
        }
    }
```

6.7.8 创建场景响应函数

当子弹射中时,爆炸产生或者一个新的敌机需要出现,这时候都需要使用到代码清单 6-26 的场景响应 spawn 函数。

代码清单 6-26

```
    p.spawnEnemyShip = function () {
        var enemy = this.enemyPool.getSprite();
        enemy.y = -enemy.getBounds().height;
        enemy.x = Utils.getRandomNumber(enemy.getBounds().width, screen_width - enemy.getBounds().width);
        this.addChild(enemy);
        this.enemies.push(enemy);
```

```
        }
    p.spawnEnemyBullet = function (enemy) {
        var bullet = this.enemyBulletPool.getSprite();
        bullet.currentAnimationFrame = 1;
        bullet.y = enemy.y;
        bullet.x = enemy.x;
        this.addChildAt(bullet, 0);
        this.enemyBullets.push(bullet);
    }
    p.spawnHeroBullet = function () {
        var bullet = this.heroBulletPool.getSprite();
        bullet.x = this.heroShip.x;
        bullet.y = this.heroShip.y - this.heroShip.getBounds().height / 2;
        this.addChildAt(bullet, 0);
        this.heroBullets.push(bullet)
    }
    p.spawnEnemyExplosion = function (x, y) {
        var explosion = this.explosionPool.getSprite();
        explosion.x = x - 45;
        explosion.y = y - 30;
        this.addChild(explosion);
        explosion.on('animationend', this.explosionComplete, this, true);
        explosion.play();
        createjs.Sound.play(game.assets.EXPLOSION);
    }
    p.explosionComplete = function (e) {
        var explosion = e.target;
        this.removeChild(explosion);
        this.explosionPool.returnSprite(explosion);
    }
```

这里用到了对象池技术，当需要时，就从创建的对象池中取出子弹、飞船和爆炸对象；敌机是从顶端随机的横坐标出现的，敌机子弹是从飞船的中心点往下进行发射；在渲染过程中敌机发生爆炸，产生此爆炸的敌机将响应 spawnEnemyExplosion 函数来在敌机的位置上用爆炸效果代替敌机，当所有的 Sprite 帧循环完毕，就产生一个监听事件，然后播放一个爆炸效果，当动画帧结束，调用 explosionComplete 来清除显示效果，返回其所在的对象池；最后的循环是调用一些检测函数，包括爆炸效果的检测等，可以使用第三方库来做爆炸碰撞检测，读者可以查阅相关书籍。

6.7.9 检测碰撞效果

太空英雄大战的碰撞效果仅仅依靠四方形的边缘检查是不够的，需要有特殊的碰撞检测方法，这里使用 EaselJS 语言编写的第三方检测碰撞工具，可以从 https://github.com/olsn/Collision-Detection-for-EaselJS 下载，然后把相应的库包放入应用程序中，通过程序进行调用。

代码清单 6-27

```
Var collision = ndgmr.checkPixelCollision(sprite1, sprite2);
```

方法返回的结果是个对象或者 false 值，对碰撞结果进行检测的代码如下。

代码清单 6-28

```
if (collision) {
//碰撞检测
}
```

6.7.10 创建检测函数

最后一个游戏的循环就是建立各种检测函数，检测函数检查各种状态并且监测事件的发生，例如爆炸的发生、游戏是否结束、关卡变化和敌方飞船是否发生炸弹等。

1. 检测敌方重建时间和攻击时间

如代码清单 6-29 所示，重建敌机或者敌机发射炸弹都是由时间进行控制的。

代码清单 6-29

```
p.checkForEnemySpawn = function (time) {
    if (time - this.enemyLastSpawnTime > this.enemySpawnWaiter) {
        this.spawnEnemyShip();
        this.enemyLastSpawnTime = time;
    }
}
p.checkForEnemyFire = function (time) {
    var enemy, i;
    var len = this.enemies.length - 1;
    for (i = len; i >= 0; i--) {
        enemy = this.enemies[i];
        if (time - enemy.lastFired > enemy.fireDelay) {
            this.spawnEnemyBullet(enemy);
            enemy.lastFired = time;
        }
    }
}
```

调用这两个函数时，将在触发事件中产生一个 time 的属性事件，当时间到达时，一个 Ticker 就被初始化：

```
this.checkForEnemySpawn(tickEvent.time);
```

这个时间参数用来计算自从上一颗子弹发射之后的时间值，如果已经到发射下一颗子弹的时间，就用 spawnEnemyBullet 进行子弹的重载，lastFired 属性值重设为 0。类似的，敌机的重载通过 enemyLastSpawnTime 来进行。

2. 检测攻击效果

如代码清单 6-30 所示，子弹和飞船、敌机和英雄都产生碰撞效果。

代码清单 6-30

```
p.checkHeroBullets = function () {
    var i, b, bullet, enemy, collision;
    for (i in this.enemies) {
        enemy = this.enemies[i];
        for (b in this.heroBullets) {
            bullet = this.heroBullets[b];
            collision = ndgmr.checkPixelCollision(enemy, bullet);
            if (collision) {
                enemy.takeDamage();
                bullet.shouldDie = true;
            }
        }
    }
}
```

```
p.checkEnemyBullets = function () {
    var b, bullet, collision;
    for (b in this.enemyBullets) {
        bullet = this.enemyBullets[b];
        collision = ndgmr.checkPixelCollision(this.heroShip, bullet);
        if (collision) {
            bullet.shouldDie = true;
            this.heroShip.takeDamage();
            this.healthMeter.takeDamage(10);
        }
    }
}
p.checkShips = function () {
    var enemy, i;
    var len = this.enemies.length - 1;
    for (i = len; i >= 0; i--) {
        enemy = this.enemies[i];
        if (enemy.y > screen_height / 2) {
            collision = ndgmr.checkPixelCollision(this.heroShip, enemy);
            if (collision) {
                this.removeChild(enemy);
                this.enemies.splice(i, 1);
                this.spawnEnemyExplosion(enemy.x, enemy.y);
                this.heroShip.shouldDie = true;
                break;
            }
        }
    }
}
```

checkHeroBullets 方法通过一个 for (i in this.enemies) 和 for (b in this.heroBullets) 嵌套循环来检测 ndgmr.checkPixelCollision(this.heroShip, bullet) 敌机是否被英雄炸弹碰到，如果碰撞产生，则调用 takeDamage() 函数，并且设置炸弹的 shouldDie 值为 true；checkEnemyBullets 以同样的方式来检测敌方炸弹是否碰撞到英雄飞船，当碰撞效果产生时，需要产生飞船毁灭 heroShip.takeDamage() 和更新健康进度条 healthMeter.takeDamage(10)。checkShips 用来检测飞船与飞船的碰撞事件，先用 if (enemy.y > screen_height / 2)来筛选出当飞船飞到屏幕下半部分的时候才开始检测，当碰撞发生，敌机不应该等待下次 tick 触发的时候再确定生死，而是应该立刻爆炸，这样做的目的是英雄的死亡可以暂停所有更新和渲染循环，英雄的生命值是由其自身的检测函数来确定的，所以需要设置 shouldDie 为 true。

3. 检测英雄死亡状态

如代码清单 6-31 所示，当英雄死亡时，checkHero 函数将被触发，该函数检测 shouldDie 参数值，当前这个值主要是由飞船碰撞飞船来改变的，checkHealth 主要通过判断 healthMeter 是否为空来设置英雄飞船的 shouldDie 值，checkHero 方法判断 shouldDie 是否为 true 就可以决定是否爆炸了；explode() 调用了一系列的爆炸效果，英雄生命值 numLives 减一，并将其从 lifeBox 中移出，最后设置 betweenLevels 为 true 来暂停游戏的动作，是否再次启动是由步骤 4 中的游戏检测来决定的。

代码清单 6-31

```
p.checkHealth = function (e) {
    if (this.healthMeter.empty) {
        this.heroShip.shouldDie = true;
```

```
        }
    }
    p.checkHero = function () {
        if (this.heroShip.shouldDie) {
            this.numLives--;
            this.heroShip.explode();
            this.lifeBox.removeLife();
            this.betweenLevels = true;
        }
    }
```

4．检测游戏

如代码清单 6-32 所示，在爆炸帧完成时，英雄飞船实例设定了一个监听事件。

代码清单 6-32

```
    p.checkGame = function (e) {
        if (this.numLives > 0) {
this.heroShip.reset();
            this.heroShip.makeInvincible(true);
            this.healthMeter.reset();
            this.betweenLevels = false;
        }
        else {
            game.score = this.scoreboard.getScore();
            this.dispose();
            this.dispatchEvent(game.GameStateEvents.GAME_OVER);
        }
    }
    p.dispose = function(){
        document.onkeydown = null;
        document.onkeyup = null;
    }
```

checkGame 方法中如果生命 numLives > 0，则英雄飞船将重新设置，makeInvincible(true)再次设置飞船的活跃度为 true，makeInvincible(true)健康进度条也重设，betweenLevels 设置为 false 则可以重新开始游戏进度；如果 numLives 已经使用完毕，游戏将结束，scoreboard.getScore()获得游戏的分数以便分数的显示，调用 dispose()处理游戏结束的一些动作，最后触发 GameStateEvents.GAME_OVER 事件。dispose 方法清除键盘事件，设置 onkeydown 和 onkeyup 为空值。

到此为止，一个完整的太空英雄大战游戏已经设计完成，具体的脚本运行文件请查看本书源代码，读者可以多次运行来体验游戏效果。

学习结果测评

非常高兴你来到这里,学完本书,希望你在学到新知识的同时,也结交到一些志同道合的朋友。接下来,你可以参加本书的学习结果测评了,成绩合格者可以申请课程结业证书,成绩优秀者将会获得额外大奖。

注意事项

- 每人只能考一次
- 每次考试的试卷都不同
- 考试开始后计时,提交试卷后停止计时
- 考试成绩会综合对错、时长等因素综合评定
- 考试结束后,会给出参考答案,以便更好地巩固知识

开始测评

(微信扫描打开试卷)

你的成绩

学习和考试系统由极客学院提供